ARCTIC REGION AND ANTARCTICA ISSUES AND RESEARCH

THE ARCTIC COUNCIL AND A CHANGING ARCTIC

ISSUES FOR THE UNITED STATES

ARCTIC REGION AND ANTARCTICA ISSUES AND RESEARCH

Additional books in this series can be found on Nova's website under the Series tab.

Additional e-books in this series can be found on Nova's website under the e-book tab.

ARCTIC REGION AND ANTARCTICA ISSUES AND RESEARCH

THE ARCTIC COUNCIL AND A CHANGING ARCTIC

ISSUES FOR THE UNITED STATES

HAYLEY JUSTINA COBB
EDITOR

nova publishers
New York

For permission to use material from this book please contact us:
Telephone 631-231-7269; Fax 631-231-8175
Web Site: http://www.novapublishers.com

Library of Congress Cataloging-in-Publication Data

ISBN: 978-1-63321-719-5

Published by Nova Science Publishers, Inc. † New York

CONTENTS

PREFACE

The Arctic Council (Council) is a voluntary intergovernmental forum for Arctic States, with involvement of indigenous organizations and other stakeholders, to address various environmental and economic issues through projects and reports targeting a variety of subjects. Recent changes in the Arctic from a warming climate, such as decreased sea ice coverage making marine areas more accessible, have increased global attention to the region's economic opportunities. This book examines the Council's organization and how it addresses environmental and economic development issues; how key U.S. agencies participate in the Council and any challenges; and agencies' actions to implement and manage voluntary Council recommendations and any challenges.

Chapter 1 – Recent changes in the Arctic from a warming climate, such as decreased sea ice coverage making marine areas more accessible, have increased global attention to the region's economic opportunities. In 1996, the eight Arctic States—Canada, Denmark, Finland, Iceland, Norway, the Russian Federation, Sweden, and the United States—formed the Arctic Council to promote cooperation on various Arctic issues with input from indigenous groups. U.S. Arctic policy highlights the importance of the Council to pursue U.S. Arctic interests. GAO was asked to examine matters related to U.S. Council participation. This report examines (1) the Council's organization and how it addresses environmental and economic development issues; (2) how key U.S. agencies participate in the Council and any challenges; and (3) agencies' actions to implement and manage voluntary Council recommendations and any challenges. GAO analyzed key documents; interviewed federal and other Arctic stakeholders; attended a Council meeting; and visited four Alaskan Arctic communities selected for their sizes and needs.

Chapter 2 – The diminishment of Arctic sea ice has led to increased human activities in the Arctic, and has heightened interest in, and concerns about, the region's future. Issues such as Arctic sovereignty claims; commercial shipping through the Arctic; Arctic oil, gas, and mineral exploration; endangered Arctic species; and increased military operations in the Arctic could cause the region in coming years to become an arena of international cooperation or competition. The United States, by virtue of Alaska, is an Arctic country and has substantial political, economic, energy, environmental, and other interests in the region. Decisions that Congress, the executive branch, foreign governments, international organizations, and commercial firms make on Arctic-related issues could significantly affect these interests.

In: The Arctic Council and a Changing Arctic
Editor: Hayley Justina Cobb

ISBN: 978-1-63321-719-5

Chapter 1

ARCTIC ISSUES: BETTER DIRECTION AND MANAGEMENT OF VOLUNTARY RECOMMENDATIONS COULD ENHANCE U.S. ARCTIC COUNCIL PARTICIPATION*

United States Government Accountability Office

WHY GAO DID THIS STUDY

Recent changes in the Arctic from a warming climate, such as decreased sea ice coverage making marine areas more accessible, have increased global attention to the region's economic opportunities. In 1996, the eight Arctic States—Canada,

Denmark, Finland, Iceland, Norway, the Russian Federation, Sweden, and the United States—formed the Arctic Council to promote cooperation on various Arctic issues with input from indigenous groups. U.S. Arctic policy highlights the importance of the Council to pursue U.S. Arctic interests. GAO was asked to examine matters related to U.S. Council participation. This report examines (1) the Council's organization and how it addresses environmental and economic development issues; (2) how key U.S. agencies participate in the Council and any challenges; and (3) agencies' actions to implement and manage voluntary Council recommendations and any challenges. GAO analyzed key documents; interviewed federal and other Arctic stakeholders; attended a Council meeting; and visited four Alaskan Arctic communities selected for their sizes and needs.

WHAT GAO RECOMMENDS

GAO recommends that State work with relevant agencies to develop a strategy identifying direction for agency Council participation and resource needs; develop a process

* This is an edited, reformatted and augmented version of a United States Government Accountability Office publication, No. GAO-14-435, dated May 2014.

to review and track progress on recommendations; and work with other Arctic States to develop guidelines for clear and prioritized recommendations. State agreed with GAO's recommendations.

WHAT GAO FOUND

The Arctic Council (Council) is a voluntary intergovernmental forum for Arctic States, with involvement of indigenous organizations and other stakeholders, to address various environmental and economic issues through projects and reports targeting a variety of subjects. The eight Arctic States guide the work of the Council through consensus decisions and rotate the chairmanship of the Council every 2 years. The United States will assume the chairmanship in 2015. The participants meet in six working groups, four task forces, and various expert groups to produce such documents as scientific assessments and guidance. For example, the Council has produced assessments of Arctic climate change impacts and shipping. As Arctic issues have emerged, the Council has expanded and broadened its work to address them. For example, since the Council's was established in 1996, the number of ongoing projects has increased from about 30 to 80.

Six key federal agencies hold leadership roles in the Arctic Council and other agencies participate through the Council's working groups and task forces. The U.S. Department of State (State) leads this participation and collaborates with the five other key agencies that lead the delegations to Council working groups—the Environmental Protection Agency, National Nuclear Security Agency, National Oceanic and Atmospheric Administration, U.S. Fish and Wildlife Service, and U.S. Global Change Research Program—as well as other federal agencies with Arctic interests. In collaborating on Council work, the agencies face challenges by not having a clear direction or specific resources for their work. For example, key agency officials said that the agencies do not have a strategy that guides and aligns their Council work. Without a clear direction or specific resources for the collaborative effort, the agencies face challenges prioritizing the work, delivering unified messages to other Arctic States, and consistently participating in the Council. GAO previously reported that agencies can enhance and sustain collaborative efforts by engaging in various practices, such as establishing joint strategies and identifying necessary resources.

Federal agencies have acted on some voluntary recommendations that the United States and other Arctic States have adopted through the Council. However, State does not review or track progress made on these actions and faces challenges implementing the voluntary recommendations. Specifically, State informally discusses the recommendations with other agencies during monthly meetings but does not have a process to review and track progress the agencies have made toward implementing them. State officials said that the agency may need to more formally assess such progress because, without such a process, State does not know the status of recommendation implementation and faces challenges planning for and prioritizing future actions to address Arctic issues. In addition, the United States—with State as the lead agency—and other Arctic States face challenges implementing the Council's broad and numerous recommendations. To address these challenges, State officials said that the Council needs to more clearly specify and prioritize recommendations, but the Council does not have guidelines for doing so. Without such guidelines, officials said the recommendations have not historically produced actions with measurable outcomes.

ABBREVIATIONS

BOEM	Bureau of Ocean Energy Management
Council	Arctic Council
EPA	Environmental Protection Agency
FWS	U.S. Fish and Wildlife Service
IMO	International Maritime Organization
NNSA	National Nuclear Security Administration
NOAA	National Oceanic and Atmospheric Administration
SAO	Senior Arctic Officials
UNCLOS	United Nations Convention on the Law of the Sea
USGCRP	U.S. Global Change Research Program

May 16, 2014

Congressional Requesters

A warming climate is changing the Arctic environment and bringing economic development opportunities to the region.[1] For example, over the 50-year period ending in 2009, average annual temperatures in Alaska have increased by 1.9 degrees Celsius (3.4 degrees Fahrenheit), more than twice the average global temperature rise for this period.[2] In addition, the coverage of summer sea ice has declined at an average rate of about 14 percent per decade since 1979, reaching its lowest point on record in 2012, according to the National Snow and Ice Data Center.[3]

As Arctic marine areas become more accessible, the region has attracted increasing global attention for its economic opportunities. For example, the U.S. Geological Survey estimated in 2008 that the Arctic has approximately 13 percent of the world's undiscovered oil resources.[4] In addition, trans-Arctic shipping routes are thousands of miles shorter than traditional routes between the Atlantic Ocean and Asia, and traffic through the Bering Strait increased from 217 vessels in 2008 to 484 vessels in 2012 and 434 vessels in 2013, according to data from the Coast Guard. However, potential increased economic development opportunities bring increased concerns about possible pollution risks from oil spills and accidents, as well as challenges responding to such incidents in the region. Amid these environmental and economic changes, many indigenous people of the Arctic struggle to maintain their traditional way of life that relies on the ecosystem for subsistence resources.[5]

These complex and interrelated Arctic issues transcend borders and increase the importance of cooperation among the Arctic States— Canada, the Kingdom of Denmark (Denmark), Finland, Iceland, Norway, the Russian Federation (Russia), Sweden, and the United States—as they address common challenges and opportunities.[6] The Arctic States began coordinating officially through the Arctic Environmental Protection Strategy adopted in 1991 and formed the Arctic Council (Council) in 1996 to promote cooperation, coordination, and interaction on Arctic issues, particularly those related to environmental protection and sustainable economic development.[7] The Council's charter expressly omitted matters related to military security. The Council's chair rotates among each of the eight Arctic States every 2 years. Canada assumed the chair in 2013, and the United States will succeed it as chair in 2015.

U.S. Arctic policy and strategy highlight the importance of working with other Arctic States through forums such as the Council to pursue U.S. Arctic interests, including protecting the environment, managing resources, involving indigenous communities and the state of Alaska, and supporting scientific research. [8] The U.S. Department of State (State) leads the U.S. participation in the Council and coordinates the related work of five other key federal agencies with a leadership role in the U.S. delegation—the Department of Commerce's National Oceanic and Atmospheric Administration (NOAA), the Department of Energy's National Nuclear Security Administration (NNSA), the Department of the Interior's U.S. Fish and Wildlife Service (FWS), the Environmental Protection Agency (EPA), and the U.S. Global Change Research Program (USGCRP)—as well as other federal agencies, such as the Coast Guard and Bureau of Ocean Energy Management (BOEM), that play various roles in the Council.[9]

We have previously reported on emerging issues and challenges for the United States in the Arctic. For example, we have reported on Alaska Native villages threatened by flooding and coastal erosion, finding that no single federal agency manages and allocates the limited assistance available to protect and relocate these villages.[10] In addition, in March 2014, we reported on U.S. Arctic maritime infrastructure challenges and the actions taken by federal, state, and local stakeholders to plan for future Arctic maritime infrastructure investments.[11] We have included a list of related products at the end of this report.

You asked us to review matters related to U.S. participation in the Council. Our objectives were to examine (1) how the Council is organized and how it addresses environmental and economic development issues; (2) how U.S. federal agencies with a key leadership role participate in the Council and the associated challenges, if any; and (3) the actions State and other federal agencies have taken to implement and manage voluntary recommendations the United States has adopted through the Council and associated challenges, if any.

To address these objectives, we analyzed documents and reports from the Council and reviewed relevant literature from government entities, academic sources, and nongovernmental organizations. We also interviewed (1) knowledgeable officials from State and other U.S. agencies with a lead role in the U.S. delegation to the Council; (2) leaders of organizations that represent Alaska Natives in the Council; and (3) Arctic stakeholders including state of Alaska officials, Alaska Natives, representatives of nongovernmental organizations with Arctic interests, and academics that we selected by referral from key federal agency officials and based on their experience participating in the Council or their knowledge of its work. In addition, we attended a Council coordination meeting in Stockholm, Sweden, and conducted site visits to the Arctic Alaskan communities of Barrow, Kivalina, Kotzebue, and Wainwright. We identified these communities through stakeholder referrals and a review of literature and reports describing the issues facing Alaskan communities, and selected them to include a range of community sizes and needs, as well as their accessibility.[12] We also analyzed Council documents from 1998 to 2013 to identify recommendations made by the Council to Arctic States. Appendix I provides additional information about our objectives, scope, and methodology.

We conducted this performance audit from November 2012 to May 2014 in accordance with generally accepted government auditing standards. Those standards require that we plan and perform the audit to obtain sufficient, appropriate evidence to provide a reasonable basis for our findings and conclusions based on our audit objectives. We believe that the evidence obtained provides a reasonable basis for our findings and conclusions based on our audit objectives.

BACKGROUND

This section discusses the origins of the Council and other international forums for addressing Arctic issues, U.S. Arctic policy, and key Arctic issues.

Arctic Council Origins and Other International Forums for Addressing Arctic Issues

The Ottawa Declaration, signed in 1996 by the eight Arctic States, formally established the Council as an intergovernmental forum with the involvement of Arctic indigenous communities and others. The eight Arctic States are Council members and have exclusive decision-making responsibilities. The Ottawa Declaration named three Arctic indigenous organizations as "Permanent Participants" and opened such status to other Arctic indigenous organizations that met specified criteria, as well as observer status for non-governmental and intergovernmental organizations that can contribute to the Council's work. The chairmanship of the Council rotates among the eight Arctic States. Appendix II lists the chairmanships since the Council's inception.

In addition to the Council, other international treaties, fora, and organizations exist that cover specific Arctic issues. For example, the United Nations Convention on the Law of the Sea (UNCLOS) treaty provides a legal framework governing the uses of the oceans and its resources. The treaty gives coastal states exclusive sovereign rights to explore and use the natural resources, including oil and gas, of their continental shelves.[13] In addition, it establishes a procedure for coastal states that intend to establish the outer limits of their continental shelf beyond 200 nautical miles, when it extends that far. The United States is the only Arctic State that is not a party to UNCLOS, but supports and observes it as customary international law and practice, as stated in the National Strategy for the Arctic Region.[14] However, without ratifying UNCLOS, the United States cannot make claims or participate in the proceedings to establish states' outer limits on their continental shelves.

Next, through an informal forum, the five Arctic States with coastal territory above the Arctic Circle—Canada, Denmark, Norway, Russia, and the United States—adopted the Ilulissat Declaration in 2008.[15] The declaration stated, in part, that these five states remained committed to the legal framework of UNCLOS and saw no need to develop a new comprehensive legal regime to govern the Arctic Ocean. In addition, at a February 2014 meeting, these five states agreed to work toward preventing unregulated commercial fishing in the central Arctic Ocean and conduct additional scientific research.

Finally, all eight Arctic States work through the International Maritime Organization (IMO), a United Nations agency, on issues related to the safety and security of shipping and the prevention of marine pollution by ships. The IMO expects to complete in 2014 a mandatory International Code of safety for ships operating in polar waters, which it terms the Polar Code.[16]

U.S. Arctic Policy

The United States has articulated its interests in the Arctic since the early 1970s. Three key documents that guide current U.S. Arctic policy and management recognize the importance of the Council in pursuing U.S. Arctic interests but do not expressly guide U.S. participation in the Council. First, the *Arctic Region Policy* issued in 2009 specifies U.S. priorities in the Arctic related to national security, international governance, international scientific cooperation, economic issues, environmental protection, and maritime transportation.[17] The policy recognized, among other things, the establishment and ongoing work of the Council and that the United States participates in the Council to promote U.S. interests in the region. Second, in May 2013 the administration issued the *National Strategy for the Arctic Region* that articulates U.S. strategic priorities in the Arctic and the federal, state, and international partners that are involved in the Arctic dialogue.[18] The strategy states that the United States participates in international forums like the Council to promote U.S. interests in the region and that the United States will pursue agreements that advance collective interests, promote shared Arctic State prosperity, protect the Arctic environment, and enhance regional security, among other things. The administration released an implementation plan in January 2014 identifying the methodology, process, and approach for executing the strategy and recognizing the importance of international engagement for successfully managing the Arctic Region.[19] Third, a 2013 report on Arctic management produced by an interagency group recognizes the expert advice the Council provides on a range of subjects and recommended that the United States continue to play an active role in the Council.[20]

Key Arctic Issues

Numerous scientific studies, assessments, and government reports have documented the widespread changes occurring in the Arctic and their local and global implications. Climate change affects many of these issues because warming in the Arctic has exceeded the warming in the rest of the world. Scientists have documented the wide-ranging effects of such warming on various aspects of the Arctic environment.[21] For example, the coverage of Arctic sea ice has steadily declined since 1979, when satellite measurements became available. As shown in Appendix II, sea ice coverage in the years from 2001 to 2013 had decreased compared to the 1979-2000 average, reaching its lowest coverage in 2012.

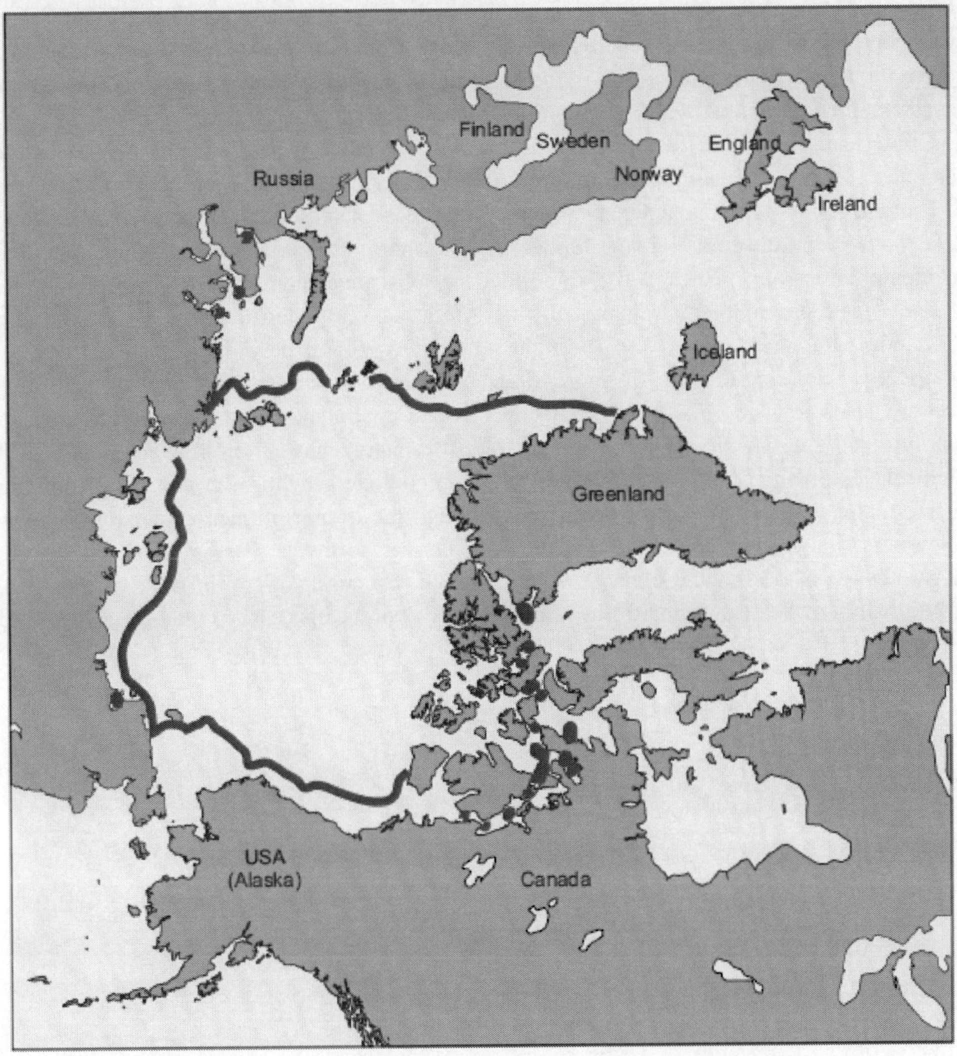

Source: National Snow and Ice Data Center.
Note: The median ice edge displays the average annual minimum position of the ice edge.

Figure 1. Median Ice Edge Coverage from 2001 to 2013.

In addition to declining sea ice coverage, rising Arctic temperatures cause thawing of the permafrost—permanently frozen ground, according to the 2009 USGCRP report.[22] A 2012 Council report states that such thawing can decrease the stability of infrastructure such as buildings, roads, pipelines, and airports, and it affects natural ecosystems through ground surface collapse and draining of lakes, among other things.[23] Also, particularly when combined with increased wave action from reduced sea ice coverage, the less stable ground becomes susceptible to coastal erosion, affecting coastal villages and community members. For example, in 2009, we reported that at least 12 Alaska Native villages decided to relocate or explore relocation options due to imminent threats from coastal erosion. Finally, as the permafrost thaws, contaminants frozen in the ice and soils can release into the air, water, and land, according to the 2012 Council report. Various Council reports state that the thawing, along with increased human activity in the region, can introduce contaminants into the Arctic ecosystem, including persistent organic pollutants and mercury.[24]

Climate change also affects the Arctic Ocean's chemistry, increasing its acidity as the ocean absorbs the increasing amounts of carbon dioxide in the atmosphere, according to the 2013 interagency Arctic management report. This report stated that such ocean acidification could have consequences for the entire Arctic marine ecosystem, including possible detrimental effects on species of fish important for subsistence communities and commercial fisheries.[25] The environmental changes in the Arctic have affected the region's people and ecosystems, but they have also brought about economic development opportunities. Specifically, according to a 2014 Congressional Research Service report, a decrease in sea ice coverage leads to a more accessible Arctic, presenting potential opportunities to expand economic activities in the region. For example, the interagency Arctic management report stated that diminishing Arctic sea ice will likely encourage growth of commercial shipping via international trans-Arctic sea routes, even though the timing of such expansion remains unclear. As shown in figure 2, these routes include the Northwest Passage that follows the northern coasts of Alaska and Canada, connecting the east coasts of North America and Asia, and the Northern Sea Route that follows the northern border of Russia, connecting Asia and Europe.[26]

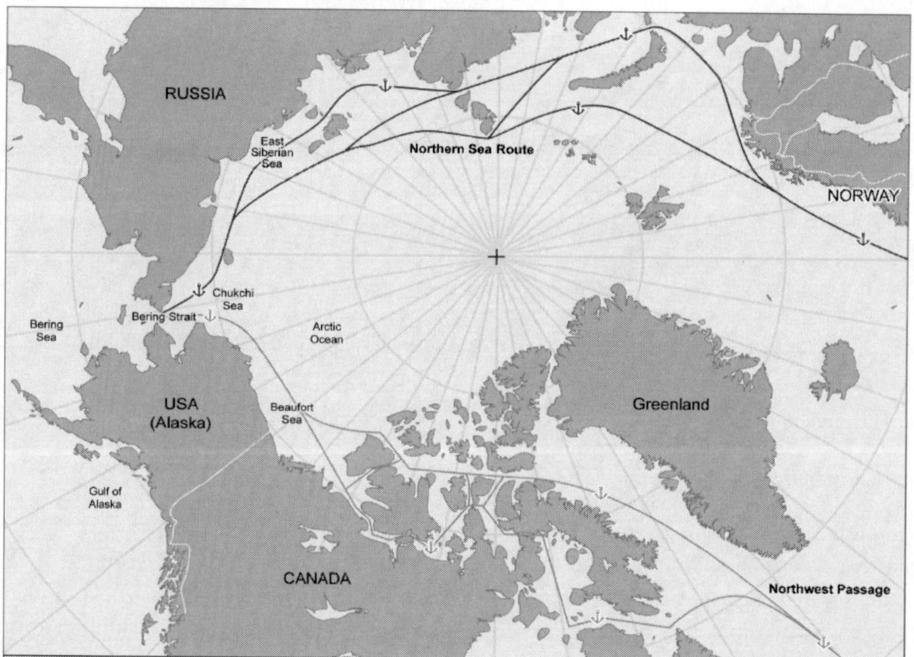

Source: Office of Naval Intelligence; GAO; and Map Resources (map).

Figure 2. Trans-Arctic Sea Routes.

Another economic opportunity arising from a more accessible Arctic includes potential development of the region's mineral, oil, and gas resources. However, the economic factors and technical challenges of operating in the Arctic environment will also affect the growth of such development, according to Department of the Interior officials and the 2014 Congressional Research Service report.

A more accessible Arctic presents economic opportunities, but such activities can lead to negative impacts. For example, as we previously reported, increases in commercial activities could increase the risk of ship collisions or oil spills, and potentially increase demand for services such as search and rescue and other maritime navigation support.[27] Furthermore, according to a 2011 Council report, increased shipping along with other human activities such as the use of diesel vehicles and residential heating can contribute to emissions of black carbon—a so called short-lived climate forcer—which can warm the climate.[28]

The Arctic Council Is a Voluntary Intergovernmental Forum for Arctic States, with the Involvement of Others, to Address Environmental and Economic Development Issues

The Council is a voluntary intergovernmental forum for the eight Arctic States, with the involvement of indigenous organizations and other stakeholders, to address environmental and economic development issues through various projects and reports.[29] Arctic States, indigenous organizations, and others serve in various roles and cooperate and coordinate through numerous Council meetings to discuss and conduct work. The Council conducts a variety of projects including scientific assessments, reports, guidance, and regional and pan-Arctic initiatives. Arctic States, indigenous organizations, and others provide financial and in-kind contributions to carry out these projects, and the Council has expanded and broadened its work to address emerging issues.

Arctic States and Others Serve in Various Council Roles

Arctic States serve as the members of the Council and make decisions by consensus among themselves. The Arctic States set the priorities for the Council, make decisions on Council matters, coordinate the Council's meetings, perform administrative functions, and finance the Council's work on a voluntary basis. Foreign ministers from Arctic states typically serve as Arctic Ministers and high-level staff from foreign ministries typically serve as Senior Arctic Officials (SAO). Staff from various agencies and ministries within Arctic State governments serve as members of the six working groups that perform the majority of the Council's technical and scientific work.

Certain Arctic indigenous organizations serve as Permanent Participants in the Council. The Ottawa Declaration named three Arctic indigenous organizations as Permanent Participants and opened such status to other Arctic indigenous organizations that met specified criteria.[30] Permanent Participants have the opportunity to participate in all Council meetings and provide input to reports and other projects but the Arctic States make Council decisions by consensus.[31] Six Permanent Participants currently represent indigenous organizations across the Arctic, and four of the six Permanent Participant organizations represent Alaska Natives (see fig. 3).

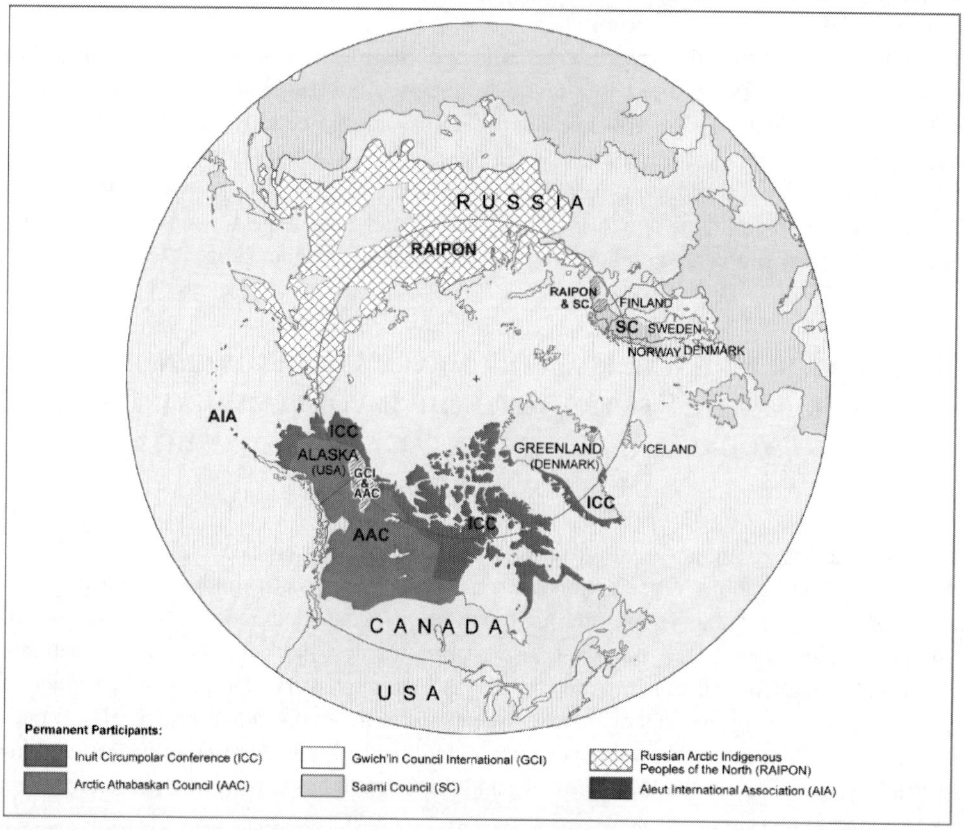

Source: Compiled by W.K. Dallman © Norwegian Polar Institute.

Figure 3. Permanent Participant Organizations of the Arctic Council.

The Arctic Council Indigenous Peoples' Secretariat provides some assistance to the Permanent Participants.[32] This Secretariat collects and communicates information about Council activities and decisions and disseminates it to the Permanent Participants.

In addition, under the Ottawa Declaration, non-Arctic states; intergovernmental, interparliamentary, global and regional organizations; and nongovernmental organizations may apply for "Observer" status. The Arctic States grant Observer status to these parties through consensus. Observer organization representatives may attend Council meetings and can actively participate in the Council's working groups and projects but Arctic States make Council decisions by consensus. As of April 2014, the states had granted 32 organizations and non-Arctic states Observer status: 12 non-Arctic states, 9 intergovernmental organizations, and 11 nongovernment organizations. Guest organizations can also attend meetings at the discretion of the chair and can contribute to certain Council work.

The Council has six permanent working groups composed of Arctic State representatives that cover a broad range of issues such as emergency response, protection of the Arctic marine environment, and the conservation of Arctic flora and fauna. In addition to the working groups, the Council can establish task forces and expert groups, also composed of Arctic State representatives, for a limited period to address a specific issue. Figure 4 identifies the Council members, Permanent Participants, Observers, working groups, and active task forces as of April 2014.

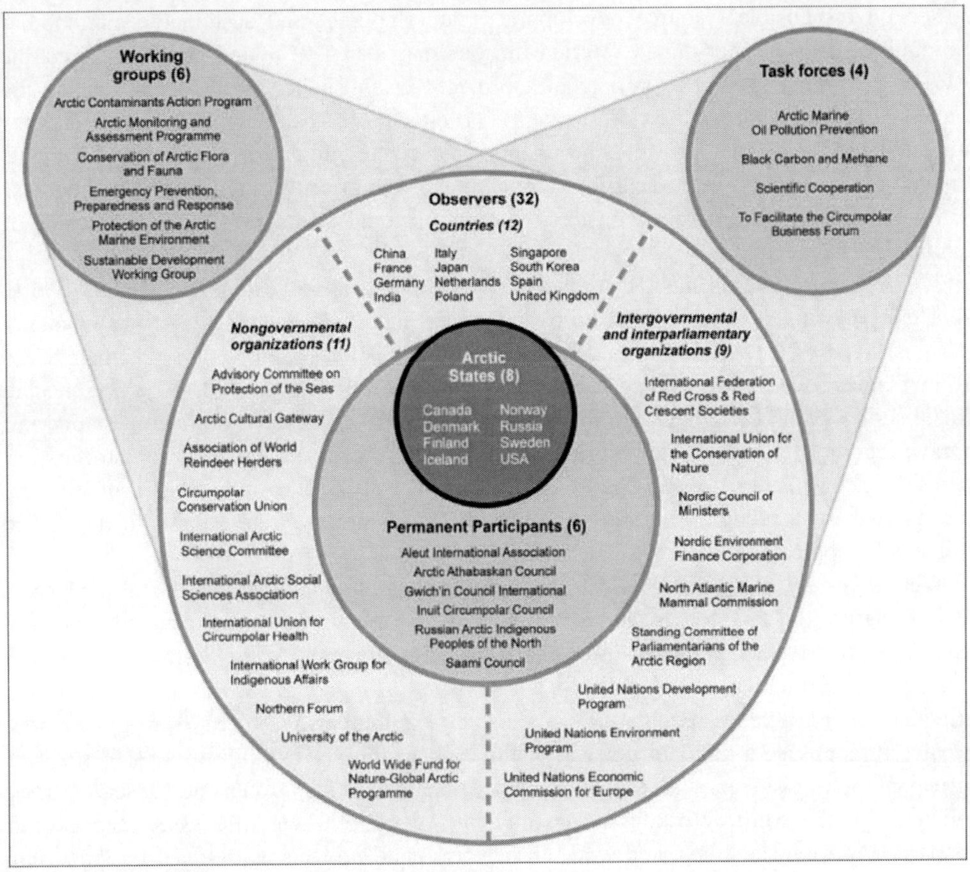

Source: GAO analysis of Arctic Council information.

Figure 4. Arctic Council Organizational Structure as of April 2014.

Arctic States and Others Cooperate and Coordinate through Numerous Arctic Council Meetings

Council participants cooperate and coordinate through numerous meetings including: (1) ministerial meetings; (2) meetings of SAOs; and (3) working group, task force, and expert group meetings.

Ministerial meetings. Arctic Ministers meet biennially to set the Council agenda for the coming 2 years.[33] The foreign ministers of the eight Arctic States typically serve as the Arctic Ministers.[34] According to senior State officials, the U.S. Secretary of State serves as the U.S.

Arctic Minister during the 2-year chairmanship and presides over the biennial Council ministerial meeting at the chairmanship's conclusion. At the ministerial meetings, the Council chair typically proposes the priorities and projects for its chairmanship. For example, during the last Council ministerial meeting held in Kiruna, Sweden, in May 2013, Canada—which assumed the chairmanship in 2013—announced that the Council would focus on "Development for the People of the North" during the next 2 years. Canada proposed projects focused on responsible resource development, safe shipping, and sustainable circumpolar communities, among other things. Arctic Ministers may also set Council priorities, direct the SAOs and working groups to initiate new projects, establish new task forces, change the Council structure, approve new Permanent Participant or Observer groups, and endorse Council reports and other projects. To document decisions, highlight the accomplishments achieved over the past 2 years, and formally announce the priorities the Council will address for the next 2 years, the Arctic Ministers issue a ministerial declaration at the conclusion of each ministerial meeting.

SAO Meetings. SAOs meet biannually to review and approve ongoing Council work. SAOs are government officials selected by the Arctic States who manage day-to-day Council activities. Most of the Arctic States designate an Arctic Ambassador to serve as their SAO, but the United States and Canada do not currently have Arctic Ambassadors.[35] A senior State official serves as the U.S. SAO. During SAO meetings, SAOs review the ongoing work and approve reports from the Council working groups, task forces, and expert groups to forward to the Arctic Ministers. Arctic States or Permanent Participants may also propose new projects at these meetings, and the SAOs decide which proposals to submit to the Arctic Ministers for approval.

Working group, task force, and expert group meetings. The Council's work takes place in the working groups, task forces, and expert groups that the Arctic States chair on a rotating basis. As such, the meetings occur in locations that span the eight Arctic States. The working groups meet biannually and the task forces and expert groups meet at times decided by the participating Arctic States. Each Arctic State selects a number of experts from government agencies to comprise a national delegation for each of the six working groups. Permanent Participants may also serve as members of a Council working group and Observers may participate in the work. Working group members carry out scientific assessments, draft reports and recommendations, and work on other Council projects as directed by the Arctic States. Each working group may establish operating guidelines and work plans. Arctic Ministers and SAOs review and approve the work plans and any revisions made to the operating guidelines for each working group.

Arctic States and Others Conduct Projects Funded through Voluntary and in-Kind Contributions

According to federal agency officials, Permanent Participant representatives, and other Arctic stakeholders, projects conducted by the Council have played a major role in advancing the global understanding of environmental and sustainable development issues in the Arctic, and enabled Council participants to discuss possible actions to address them. These projects include scientific assessments, guidelines, recommendations, best practices, and other initiatives. Some key Council assessments and guidance include the following:

- *The Arctic Climate Impact Assessment,* 2004: The Arctic Climate Impact Assessment describes ongoing impacts of climate change in the Arctic and their consequences such as rising temperatures; loss of sea ice; unprecedented melting of the Greenland ice sheet; and many impacts on ecosystems, animals, and people.[36] This assessment was the first comprehensively researched, fully referenced, and independently reviewed evaluation of climate change and impacts in the Arctic.
- *The Arctic Marine Shipping Assessment,* 2009: The Arctic Marine Shipping Assessment is the first comprehensive analysis of current and potential future Arctic shipping activity.[37] It provides 17 recommendations, which are still being implemented, related to enhancing Arctic marine safety, protecting Arctic people and the environment, and building Arctic marine infrastructure.
- *Community Based Monitoring Handbook,* 2010: Federal officials described this *Handbook* as the best manual available for researchers to use when interacting with local communities in the Arctic. [38] In particular, they said that the handbook's positive examples of engaging with indigenous communities helped guide many researchers that may have had limited previous contact with indigenous communities.

In addition to assessments and guidance, the Council has undertaken regional and pan-Arctic initiatives that produced concrete deliverables and, in some cases, led to measurable outcomes. For example, the Arctic Contaminants Action Program project on obsolete pesticides began in 2001 to reduce environmental releases of outdated pesticides found in 11 Northern territories of Russia. Through this project, working group members report they have inventoried, repackaged, and stored nearly 7,000 tons of outdated pesticides. [39] In another example, the Arctic Maritime and Aviation Transportation Infrastructure Initiative carried out by the Sustainable Development Working Group developed a Web-based searchable map of Arctic infrastructure to better identify capacity and capability needs for future infrastructure development.[40]

The Council does not have a central source of funding for projects, and all funding is voluntary. Specifically, each working group relies on its own mechanisms to fund projects. State officials said that Arctic States provide nearly all funding and in-kind contributions. The officials also said that Permanent Participants sometimes provide small amounts of funding, but their contributions to the Council's work are almost entirely in-kind. Financial support from Observers cannot exceed that of the Arctic States, unless the SAOs decide otherwise by consensus. In addition, some working groups seek supplemental financing from international financial institutions to carry out some projects. According to some Council stakeholders, working groups may compete for project funding because different working groups might simultaneously have several applications for financing under review at the same international financial institution without any coordination. This arrangement places responsibility for the prioritization of projects in the hands of these external institutions rather than with the Council itself. Some federal agency officials also said that the voluntary funding structure makes the working groups and their projects vulnerable to political forces because Arctic States fund projects that reflect their own priorities. Further, federal officials said that these projects may not always reflect Council priorities and that the projects undertaken may not always be the most important or the most strategic.

The Arctic Council Has Expanded and Broadened Its Work to Address Emerging Issues

As the opportunities and challenges facing the Arctic have grown in volume and complexity, the Council has increased its workload and created new task forces to address emerging issues. According to federal officials, when the Council was established in 1996, it had about 30 ongoing projects. Over the years, that number has increased, and State officials now estimate that the Council has about 80 ongoing projects. In addition, the Council has created a number of task forces to supplement the work of the working groups. The first task force was initiated at the 2009 Norwegian Ministerial meeting to negotiate a search and rescue agreement that was delivered at the 2011 Danish Ministerial meeting. The Council currently has a total of four task forces to address emerging issues during the 2013 to 2015 Canadian chairmanship. Some federal officials said that task forces address specific cross-cutting issues that require high-level discussion from decision makers. Two task forces, one during the Danish chairmanship from 2009 to 2011 and one during the Swedish chairmanship from 2011 to 2013, resulted in two multilateral agreements among the eight Arctic States. However, according to federal officials, the expanded use of task forces sometimes pulls experts away from the core working groups. Further, one official said that if these task forces respond to new priority issues, the Council may want to revisit the structure and scope of the existing working groups and incorporate the emerging issues into the mandate of the working groups.

The Canadian chairmanship has further broadened the Council's work to address emerging issues. For example, Canada's plan for its chairmanship includes facilitating the creation of a business forum to foster sustainable economic development in the Arctic and provide opportunities for businesses to engage with the Council.[41] Some stakeholders said that, prior to this initiative, the Council did not have mechanisms to involve industry and businesses, but some industry and business experts participate on a meeting- or project-specific basis. According to an Arctic stakeholder, business and industry have an interest in Council activities because the discussions and decisions made at the Council can directly affect their work. Moreover, stakeholders said that industry and business representatives provide valuable knowledge, expertise, and equipment for oil spill and other emergency prevention, preparedness, and response.

As the Council's workload has expanded, the Council has facilitated and integrated its operations. For example, to enhance administrative support and continuity of its operations, the Council established a small, voluntarily funded administrative Secretariat in Tromsø, Norway, in January 2013. Prior to this development, the Arctic State holding the chairmanship performed the Secretariat functions. The Secretariat, which had a staff of seven at the time of our review, maintains and updates the Council's website; provides a repository for information; and facilitates information sharing among the Council working groups, task forces, and expert groups. Federal officials said that the Secretariat will enhance the institutional memory of the Council, increase information sharing, help organize activities across the working groups, and make the work of the Council more manageable and transparent. Some Arctic stakeholders said that the Secretariat will also help strengthen relationships with other organizations by providing up-to-date information regarding the Council's projects.

In addition, federal officials said that the Council has adopted measures aimed at increasing communication and cooperation among the various Council working groups. For example, the Council encourages representatives from working groups to attend one another's meetings. Federal officials said that in early years of the Council, the Council working groups operated independently with little interaction. According to some Arctic stakeholders, this limited communication led to some competition for projects among the Council working groups. In addition, the Council issued a strategy in 2012 to improve internal communication, facilitate cooperation, and avoid overlap in Council work. Among other things, the strategy recommends that working groups prepare 2-page summaries following all meetings and distribute the summaries to the SAOs, Secretariat, working groups, and chairmanship to quickly inform them of meeting outcomes.

SIX KEY U.S. AGENCIES PARTICIPATE IN THE COUNCIL WITH LEADERSHIP ROLES AND FACE CHALLENGES BY NOT HAVING A CLEAR DIRECTION OR SPECIFIC RESOURCES FOR THE WORK

Six federal agencies with leadership roles participate in the Council mainly by heading U.S. delegations to working groups and face challenges in doing so. Other federal agencies participate in Council working groups and lead U.S. participation in some task forces. Numerous agencies with Arctic interests collaborate through an informal interagency group to conduct Council work but face challenges because of not having a clear direction and specific resources for their participation in the Council. Finally, some mechanisms exist to consider Alaska's and Alaska Natives' interests in federal agencies' Council work, but agencies face additional challenges considering these interests.

Six Key Agencies Hold Leadership Roles in the Council, and Other Agencies Participate Through Working Groups and Task Forces

Six key agencies hold leadership roles as head of the U.S. delegations to the Council's working groups, and other agencies lead the U.S. efforts on Council task forces (see fig. 5).

As one of the six key agencies, State leads federal agency efforts and participates in the Council in several capacities. For example, in addition to the Secretary of State's role as the U.S. Arctic Minister, State leads the U.S. chairmanship of the Council, which first occurred from 1998 to 2000 and will occur again in 2015.[42] A State official also serves as the U.S. SAO and leads U.S. participation in the Sustainable Development Working Group. State officials said they consult with federal agencies to select U.S. heads of delegation and U.S. representatives to the Council's working groups and task forces. For example, State asked USGCRP to head the U.S. delegation to the Arctic Monitoring and Assessment Programme working group and EPA to co-chair a task force on Short-Lived Climate Forcers with Norway and Sweden that produced a report on black carbon, among other products.[43] Some of these agencies provide varying amounts of funding to the working groups and task forces in which they participate.[44]

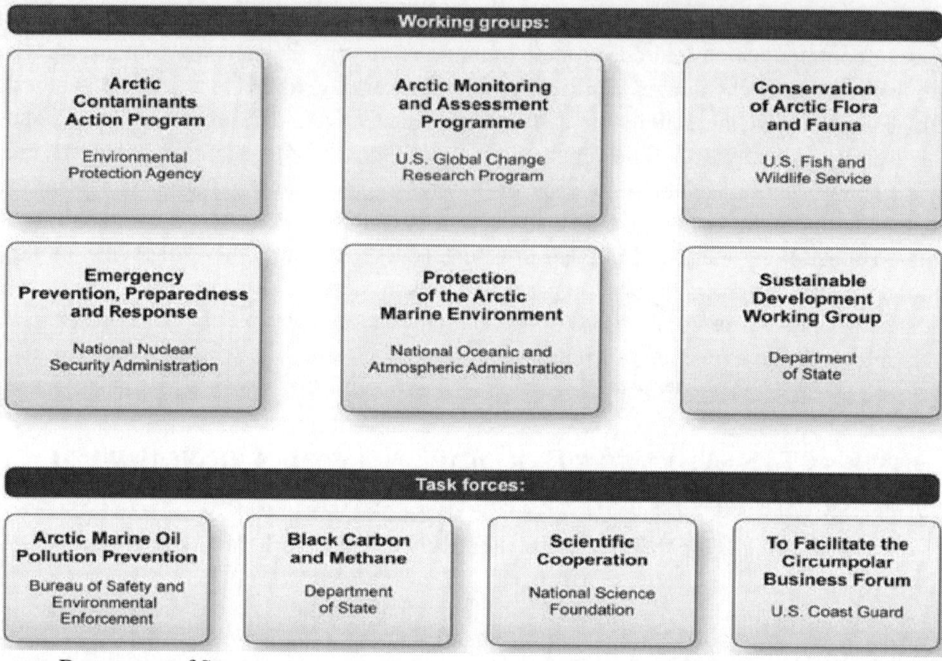

Source: Department of State.

Figure 5. U.S. Agency Leads for Arctic Council Working Groups and Task Forces as of March 2014.

In addition to agencies with leadership roles, at least 15 agencies with Arctic interests participate in the Council's working groups and task forces (see app. IV for a list of agencies participating in working groups and task forces as of January 2014). For example, a BOEM official participates in the Arctic Monitoring and Assessment Programme working group's effort to produce a three-volume oil and gas assessment, which discusses the potential effects of oil and gas activities in the Arctic. Furthermore, Coast Guard officials said they formerly served as the head of the U.S. delegations for the Task Forces on Search and Rescue and on Arctic Marine Oil Pollution Preparedness and Response, and authored various chapters of the Arctic Marine Shipping Assessment.

Most officials we interviewed from key agencies said their participation in the Council enhances their ability to leverage resources, fulfill their missions, learn best practices, or further develop relations with other Arctic States. For example, an FWS official said the work of the Council enables the agency to leverage resources to accomplish Council projects that align with its mission, and provides access to other Arctic States with common interests. In addition, a NOAA official said the Council helps the agency build partnerships that have provided opportunities crucial for conducting its work, such as obtaining access to other countries' exclusive economic zones to conduct scientific research.[45] A Coast Guard official said the participation in the Council facilitated valuable interactions with international partners to sponsor research, identify best practices, and support development of shipping standards at the IMO and other standards bodies. Coast Guard officials also said working through the Council particularly enhanced bi-lateral relations with Russian and Canadian officials, supporting related initiatives outside the Council such as updating joint contingency plans between the United States and Canada to manage a cross-border oil spill, among other things.

Agencies Collaborate on Arctic Council Work and Face Challenges by Not Having a Clear Direction or Specific Resources for the Work

Within the United States, federal agency officials who participate in the Council collaborate primarily through the Arctic Policy Group, an informal interagency group led by State but is not part of the Council itself. According to State officials, this group discusses Council issues and shares Arctic-related information, among other things.[46] State convenes the group monthly and uses it to circulate information and identify agency officials to work on Council projects. According to a National Security Decision Memorandum and State officials, the Arctic Policy Group oversees implementation of U.S. Arctic policy but does not make policy decisions. As of December 2013, this group consisted of officials from at least 39 federal agencies and the state of Alaska Governor's and Lieutenant Governor's offices (see app. V for a list of agencies participating in the group). In addition to the Arctic Policy Group, EPA officials said the six U.S. heads of delegation collaborate through informal meetings, which generally occur quarterly and involve discussion of current Council work and priorities.

Several agency officials we spoke with said that the Arctic Policy Group helps them collaborate on Council work, but some officials said agencies face challenges participating on the Council. First, some key agency officials and Arctic stakeholders said collaborating agencies face a key challenge by not having a clear direction to guide and align their Council work. Specifically, some agency officials and these stakeholders said that the six agencies have not developed a strategy that guides agency collaboration for Council participation and aligns the agencies' work. For example, a USGCRP official said, aside from State's overall role as the lead U.S. agency, other federal agencies have limited ability to influence one another and have no strategy to guide their Council work. The U.S. government and various agencies have Arctic strategies, but these documents do not articulate an overall approach for agencies to follow in Council participation. State officials said these strategies do not guide agency participation in Council activities instead, they reflect work conducted in the Council.

In the absence of clear direction that helps align collaborating agencies' Council work, some stakeholders said the United States faces challenges prioritizing the work and delivering unified messages to other Arctic States. For example, an NNSA official said that U.S. agencies do not have a strategy applied across the agencies, but agency officials discuss and agree upon projects within the Arctic Policy Group as they arise. In addition, one stakeholder who participated in the Council said different agencies have delivered inconsistent messages when providing comments on various working group reports. The stakeholder said at times agency officials provide comments on reports but do not do so in an official capacity. The stakeholder also said that establishing Council priorities could unify U.S. agencies' voices.

In addition to not having a clear direction for Council participation, officials from most of the six key agencies said they face challenges sustaining their collaborative Council efforts because they do not have specific resources for their Council work. In particular, agency officials said federal funding and staff time for Council activities come from general program budgets, which creates challenges when Council work must compete with other agency priorities for funding. For example, NOAA officials said their agency focuses mainly on domestic issues, which limits funding for international activities such as Council work even though this work relates to NOAA's mission. State officials also said the United States does not have specific funding for Council work because it had not formerly prioritized Council participation. In addition, agency officials said they do not identify their in-kind time spent on Council activities, primarily because of challenges differentiating Council tasks from their other duties.

Some agency officials said that without specified resources for their Council work, they face challenges implementing agreed-upon actions and committing to Council activities. For example, EPA officials said the absence of funding specific to Council projects limits their ability to participate. In particular, EPA officials said Council ministers approved projects on mercury contamination, but these projects had not occurred at the time of our review because of limited funding from Arctic States, including the United States. In another case, according to a FWS official, the FWS official who leads the U.S. delegation for the Conservation of Arctic Flora and Fauna working group was not able to attend in-person two working group meetings in 2012 and 2013, although FWS officials attended these meetings in previous years.[47] Some stakeholders said inconsistent U.S. participation concerns other Arctic States, which may cause some loss of U.S. influence in the Council. Furthermore, State officials said the absence of specified funding for Council work makes it challenging to manage the increased workload. For example, State officials said the agency had two full-time employees dedicated to Council work, as of February 2014. Various agency officials and stakeholders said that State does not have enough staff necessary to carry out the agency's increasing Council responsibilities. State officials said they hoped to identify additional funding for staff dedicated to Council work and were developing a plan at the time of our review to identify their anticipated staff needs for the U.S. chairmanship in 2015.

State officials said that developing a Council strategy, in collaboration with other agencies that participate in Council activities, which outlines a clear direction for participation and identifies resources might help sustain the collaborative effort to address these challenges and support the United States' role as chair in 2015. We previously reported on key practices for enhancing and sustaining collaboration among federal agencies.[48] These practices include collaborating agencies (1) establishing mutually reinforcing or joint strategies to help align activities and resources for the collaborative effort and (2) identifying and addressing needs by leveraging resources, which entails, among other things, identifying human and financial resources needed to initiate or sustain a collaborative effort. Without engaging in these practices, the collaborating agencies that hold leadership roles may continue to face challenges prioritizing the work, delivering unified messages, and consistently participating in the Council.

Agencies Consider Alaska's and Alaska Natives' Interests in Their Arctic Council Work, but Challenges Exist with These Efforts

Key federal agency officials we spoke with said they consider the state of Alaska and Alaska Natives' interests though various mechanisms, including the Council and the Alaska Arctic Council ad hoc working group, but they said challenges exist in doing so. In addition, key documents articulate the federal government's intent to consider the interests of Arctic stakeholders. For example, the 2013 *National Strategy for the Arctic Region* includes fostering partnerships with the state of Alaska and consulting and coordinating with Alaska Natives and a 2013 interagency report to the President prioritizes partnering with them to implement the national strategy.[49] However, officials from the state of Alaska and some Alaska Natives we spoke with had concerns about the extent to which federal agencies and their Permanent Participant representatives view them as partners and consider their interests since the United States is an Arctic nation because of Alaska.

Regarding the state of Alaska, federal agency officials that we interviewed mainly consider its interests in their Council work through the Arctic Policy Group, direct interactions on the U.S. delegation to the Council, and the Alaska Arctic Council ad hoc working group consisting of interested parties in Alaska. Most of the key federal agency and state of Alaska officials we interviewed said the Arctic Policy Group is the primary mechanism for considering the state of Alaska's interests. An Alaska official cited the Arctic Policy Group as a helpful way to incorporate input from the Alaska region. For example, in February 2012, an Alaska state official presented the Arctic Policy Group with a state legislative-led task force's efforts to identify opportunities for addressing the economic and environmental effects of diminishing sea ice on Alaska, according to State officials.[50]

In addition, state of Alaska officials said they work directly with federal agencies by participating in Council working groups and projects. For example, regarding the upcoming U.S. chairmanship of the Council, Alaska officials sent State a list of four priority themes the state of Alaska would like considered when the United States assumes the chair in 2015, such as creating jobs and economic opportunities for Arctic residents and ensuring safe, secure, and reliable shipping.[51] In addition, state of Alaska officials said they participated on the Council's Task Force on Arctic Marine Oil Pollution Preparedness and Response and the Arctic Maritime and Aviation Transportation Infrastructure Initiative.[52] Finally, Alaska officials said they also host a bimonthly Alaska Arctic Council ad hoc working group teleconference, which involves coordination with State officials, Alaska-based Permanent Participants, the Alaska Congressional Delegation, and others to facilitate communication on Council issues.

Alaska state officials we spoke with said they faced three main challenges while working with the federal agencies to have them consider the state's interests in the agencies' Council work. First, Alaska officials said the state's Arctic priorities sometimes differ from those of the federal agencies, making it challenging for the state to have its perspectives incorporated into agencies' Council work. For example, Alaska officials said they tried to have their interests reflected in a report by the Council's Ecosystem Based Management expert group, but could not do so because the U.S. delegation working on the report had different priorities than the state of Alaska. Second, state of Alaska officials said State does not provide enough information on the process used to make policy decisions regarding the Council. Without this information, Alaska state officials said it is difficult to know which federal agency they should contact to provide their input and when to do so. In particular, Alaska officials said that federal agencies leading U.S. participation in Council working groups have a considerable amount of autonomy to direct participation and ability to affect project outcomes, which may not align with Alaska's priorities.[53] Third, Alaska officials said the state had limited resources to engage with the federal agencies and participate in Council projects. Because of limited information on the process used for federal decision making regarding the Council and resources for Alaska's participation, Alaska state officials said they could not participate as fully as desired to ensure that the Council addresses their interests. Overall, an Alaska state official said that increasing collaboration between the state and federal government would help address some of these challenges.

With respect to incorporating the perspectives of Alaska Natives, some federal agency officials leading U.S. participation in the Council said they consider Alaska Natives' interests through engagement with the four Permanent Participant groups that represent Alaska Natives. For example, State officials said they primarily consult with Permanent Participant representatives prior to SAO meetings on an as-needed basis to discuss items on upcoming meeting agendas. Officials from other agencies said they work with the Permanent Participant representatives during Council meetings. For example, a USGCRP official said he engages with the Inuit Circumpolar Council through the Arctic Monitoring and Assessment Programme working group.[54] In addition, a Permanent Participant representative said his group regularly talks with the U.S. head of delegation to the Conservation of Arctic Flora and Fauna working group, an FWS official based in Alaska. However, this Permanent Participant representative said he interacts less frequently with federal officials located outside of Alaska, such as the U.S. head of delegation to the Sustainable Development Working Group, a State official based in Washington, D.C.

Various agencies and Alaska-based Permanent Participant representatives we spoke with said Permanent Participants' limited financial resources and number of staff for Council participation created challenges when considering Alaska Native interests in agencies' Council work. One Permanent Participant representative said over the years, increases in the number of Council activities, projects, and task forces in turn increased the burden on the Permanent Participants, particularly for travel to meetings. To help address these issues, State provides annual travel funding to the four Alaska-based Permanent Participant groups. [55] Without State funding, the groups would have a more difficult time maintaining their Council participation, according to a Permanent Participant representative. In addition to travel challenges, the increasing number of Council projects challenges Permanent Participants' capacity to keep up with the workload. For example, a former Permanent Participant representative said that effective involvement requires participation in activities taking place between SAO meetings, such as spending time reviewing draft documents. This can pose challenges for some Permanent Participants because of a limited number of staff.

Specifically, some of the Alaska-based Permanent Participant groups have no permanent staff, and others have only two to three staff.

Our site visits to Alaska Native villages reinforced some of these challenges. Alaska Natives we interviewed from Barrow, Kivalina, Kotzebue, and Wainwright said some of their primary environmental and economic interests included the following:

- ensuring sustainable economic development;
- preventing pollution from oil spills and other hazardous materials;
- reducing the effects of increased shipping traffic on the wildlife they hunt and rely on for food, such as whale, caribou, walrus, seal, and salmon;
- addressing the adverse effects of climate change, such as sea ice retreat and coastal erosion;
- reducing the effects of rising fuel prices; and
- preparing for increased visitors from cruise ships and any related incidents.

One community leader said oil and gas development would provide jobs and address infrastructure gaps, but he would like this development conducted responsibly. Other community leaders said the changes in the climate, such as declining sea ice, make it more difficult for Alaska Natives to hunt the animals they primarily rely on for food and require some villages to construct sea walls or relocate due to rising seas and coastal erosion.[56] See figure 6, an interactive graphic, for information on each site visit location (see app. VI for the full text of this graphic).

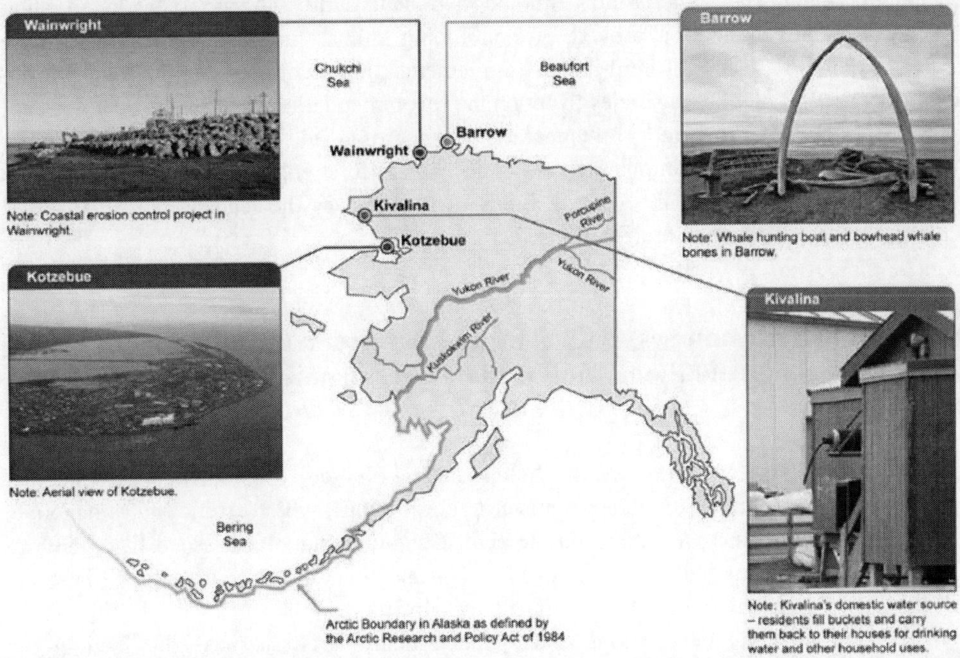

Source: Department of the Interior; GAO (analysis and photos); U.S. Census Bureau; Cities of Barrow and Kotzebue; Government of Canada; Maniilaq Association; Map Resources (map); NANA Regional Corporation; North Slope and Northwest Arctic Boroughs; State of Alaska.

Figure 6. Arctic Alaska Site Visits. See Appendix VI for more information.

Regarding representation of Alaska Native interests by the Permanent Participant groups, some Alaska Native community leaders we interviewed said they coordinate with their representative, and others expressed concern that their Permanent Participant group did not adequately represent them. For example, some Alaska Native community leaders said their Permanent Participant representative visited and worked with their community in various capacities, such as on a Council food security project. However, another Alaska Native community leader said the Permanent Participants represented too broad of a population, and he would like to have a Permanent Participant representative from his borough—a unit of government similar to a county. This community leader also said a Permanent Participant representative had never visited the community. Other Alaska Native community leaders we spoke with said they were not aware of the Council or how it represented their interests.

AGENCIES HAVE ACTED ON SOME ARCTIC COUNCIL VOLUNTARY RECOMMENDATIONS BUT STATE DOES NOT TRACK ACTIONS AND FACES CHALLENGES IMPLEMENTING THEM

Federal agencies have taken various actions on some voluntary recommendations that the United States and other Arctic States have adopted through the Council, but State does not have a process to review and track progress made on these actions and faces challenges implementing them. The United States endorses actions to address Arctic issues by adopting voluntary recommendations approved in nonbinding ministerial declarations and, more recently, signing two formal multilateral agreements that address specific high-priority issues.[57] State relies on other agencies to act on the recommendations, but the agency does not review or track their overall implementation. According to State officials and other stakeholders, the recommendations provide a mechanism to strengthen the Council's ability to address Arctic issues, but the United States—with State as the lead agency—and other Arctic States face challenges implementing them.

The United States Endorses Actions to Address Arctic Issues through Voluntary Recommendations and Formal Agreements and Has Taken Some Related Actions

The United States and other Arctic States adopt voluntary recommendations that the Council makes in nonbinding biennial ministerial declarations, which Arctic States negotiate and approve by consensus. We analyzed the eight Council declarations issued from 1998 to 2013 and consulted with State officials to identify statements in the declarations that State considers as recommendations to Arctic States. In total, State officials identified 39 recommendations in the declarations that spanned many Arctic issues addressed in the Council including the marine environment, biodiversity, climate change, environmental protection, emergency prevention, and human health and development (see app. VII for the list of recommendations that State officials identified). In general, the recommendations either identified broad actions associated with various Arctic issues or endorsed numerous recommendations made in reports and assessments produced by working groups and task forces.[58]

State relies on other agencies to act on the recommendations, and agencies have taken some actions in response. For example, several U.S. agencies have actively contributed to a monitoring initiative endorsed in the 2006 Salekhard Declaration.[59] Specifically, the declaration encouraged the Arctic States to contribute to the Circumpolar Biodiversity Monitoring Program, an international network of scientists, governments, indigenous organizations, and conservation groups working to integrate efforts to monitor the Arctic's living resources. In response, U.S. agencies have contributed to the program through the Bureau of Land Management co-chairing the effort with the North Slope Science Initiative, an intergovernmental effort to improve collection and dissemination of ecosystem information in the U.S. Arctic, and the Arctic Research Centre in Denmark.[60]

Other actions recommended by the Council were already under way in the United States. For example, in 2009, the Council recommended that the Arctic States apply guidelines developed by the Council for offshore oil and gas development.[61] According to a BOEM official, these guidelines reflected some practices already used in the United States, such as conducting environmental studies to establish baseline information for a region included in an oil and gas lease sale. In another example, the Council declarations in 2011 and 2013 encouraged Arctic States to implement recommendations in Council reports on short-lived climate forcers, including black carbon.[62] According to State officials, the United States has already implemented some actions aimed at reducing black carbon pollution. These actions include developing standards under the Clean Air Act to reduce emissions of particulate matter, of which black carbon is a component, as well as implementing voluntary EPA programs to retrofit diesel vehicles, the largest source of black carbon emissions.

In addition to voluntary recommendations adopted through the Council, starting in 2011, the United States signed multilateral agreements with the other Arctic States on search and rescue and oil spill preparedness and response that were developed by Council task forces.[63] In the 2011 *Agreement on Cooperation on Aeronautical and Maritime Search and Rescue in the Arctic* (Arctic Search and Rescue Agreement), the Arctic States committed to search and rescue actions, such as

- promoting the establishment, operation, and maintenance of an adequate and effective search and rescue capability within an area specified in the agreement;
- ensuring that assistance is provided to anyone in distress regardless of the nationality or status of such a person or circumstances in which that person is found;
- exchanging information that may serve to improve the effectiveness of search and rescue operations; and
- meeting with the other parties on a regular basis to consider and resolve issues regarding practical cooperation.[64]

In the 2013 *Agreement on Cooperation on Marine Oil Pollution Preparedness and Response in the Arctic* (Arctic Oil Pollution Preparedness and Response Agreement), Arctic States committed to oil spill preparedness and response actions, such as

- maintaining a system for responding promptly and effectively to oil pollution incidents;
- establishing, as appropriate, a minimum level of prepositioned oil spill combating equipment;
- establishing, as appropriate, a program of exercises for oil pollution response organizations and training of relevant personnel;
- establishing, as appropriate, plans and communications capabilities for responding to an oil pollution incident;
- promoting cooperation and coordination by endeavoring to carry out joint exercises and training; and
- developing and maintaining, with the other parties to the agreement, a set of operational guidelines to assist in the agreement's implementation but these guidelines will not be legally binding.[65]

The Coast Guard acts as the lead agency for implementing the two agreements and has taken or planned various actions in response. These actions include the following:

- **Arctic Search and Rescue Agreement.** According to Coast Guard officials, the Arctic Search and Rescue Agreement solidified partnerships with neighboring countries and formalized an existing understanding but did not formally commit the United States to allocate additional resources to search and rescue in the Arctic. Since the agreement was signed, the United States has organized some high-level meetings and committed to joint training exercises. Even while meeting the intent of the agreement, the absence of a permanent physical presence in the region limits the U.S. government's ability to respond to vessels or persons in distress in the Arctic, according to Coast Guard and State officials.
- **Arctic Oil Pollution Preparedness and Response Agreement.** The Coast Guard has not dedicated any additional resources to implement this agreement but plans to participate in the first exercise under the agreement in 2014 with the other Arctic States, according to Coast Guard officials. An oil and gas industry stakeholder said that the Coast Guard could face challenges implementing the agreement in the event of an oil spill incident because the government does not own adequate equipment to respond to an emergency. This stakeholder said that oil and shipping companies own approximately 90 percent of the emergency response equipment in the Arctic. However, Coast Guard officials said the agency plays an oversight role in maritime incident response and that industry must provide any response resources.

State Does Not Track Agencies' Progress Toward Implementing Recommendations

State does not review or track agencies' overall progress toward implementing the recommendations the United States has adopted through the Council. According to State officials, agencies discuss the implementation of recommendations informally during monthly interagency Arctic Policy Group meetings, but State does not formally follow up with the agencies to identify their progress toward implementation. Several senior agency officials said that historically, the United States has not made a concerted effort to follow up on recommendations because the Council had not been a high priority for State. However, the officials said that State has increased the priority of the Council recommendations as the United States prepares to assume the Council chair in 2015.

Senior State officials said that they may need to more formally focus on follow-up and provide an overall assessment of progress made by agencies toward implementing recommendations. According to the officials, developing a process for U.S. agencies to review and track implementation progress would increase accountability across the government, enhance coordination across agencies, and avoid overlap. However, the officials also said that they would have to weigh the benefits of such a process against the resources it would require for maintenance.

Without such a process to review and track progress toward implementing Council recommendations, State officials said that State and other key agencies do not know the status of the recommendations and could face challenges planning for and prioritizing actions to address Arctic issues. Specifically, State officials said that the agencies use the same resources to implement recommendations as they do for their other Council work, so adopting new actions reduces the resources available to take on the increasing workload of the Council working groups, task forces, and other efforts. The officials said they need to know the capacity of agencies to implement new recommendations as State prepares for the Council chairmanship through steps such as developing new project proposals and determining priorities.

Broad and Numerous Recommendations Create Implementation Challenges

Some federal officials and stakeholders directly involved in Council work said that the United States—with State as the lead agency—and other Arctic States at times face challenges implementing the broad recommendations made in the declarations and the numerous recommendations made in the working group and task force reports. First, according to the officials, the ministerial declarations do not always clearly specify measurable actions that the Council recommends Arctic States take. For example, the broad recommendations included such statements as, the Arctic States "encourage actions to reduce the risks of radioactivity in the Arctic" and "urge implementation of early actions where possible on methane and other relevant short-lived climate forcers" in the 2000 and 2009 declarations, respectively.[66]

In addition to the broad declaration recommendations, our analysis found and State officials said the working group and task force recommendations endorsed by the declaration statements are too numerous for Arctic States to fully implement. State officials said that a complete list of these recommendations does not exist, but our analysis identified several instances in which working group and task force reports and assessments had from 10 to 20 recommendations directed to the Arctic States.[67] For example, a 2004 Council climate change report included 16 recommendations related to climate change, and a 2009 Council report on Arctic pollutants made 18 related recommendations.[68] In this context, State officials said that the working groups produced too many recommendations to the Arctic States without specifying priorities, making it difficult for the States to implement all of them or focus on priority issues.

According to senior State officials, the Council does not have guidelines for developing clear and prioritized recommendations for the Arctic States. The officials said that not having such guidelines has contributed to the Council adopting broad and numerous recommendations, resulting in implementation challenges. State officials said that Arctic States intend for their governments to implement the recommendations, but act on them voluntarily and must balance recommendation implementation with competing needs and resources for the increasing volume of Council work.[69] Recognizing some of these challenges, the Council has started to take a more strategic approach to adopting recommendations than it had in the past, placing a greater focus on outcomes, according to some federal officials and stakeholders. Starting in 2006 the Council began developing a more formal process for reviewing and negotiating recommendations. However, challenges implementing the broad and numerous recommendations remain and State officials said that a need exists for the Council to more clearly specify actions with measurable outcomes and prioritize the Council recommendations to better ensure their implementation.

State officials and stakeholders said that the Council has produced valuable work but, partially due to these implementation challenges, the recommendations have not historically produced actions leading to measurable outcomes in addressing Arctic issues within the

Council's purview. The officials said that the recommendations, if implemented, provide a means to strengthen the Council's ability to address Arctic issues, which represents a goal of the Council and the United States. Specifically, Council declarations have underscored the importance of such strengthening, and U.S. Arctic policy states that the United States seeks to strengthen institutions for cooperation among the eight Arctic States, such as the Council. When the United States assumes the chairmanship in 2015, it will have an opportunity to guide and more effectively advance the Council's efforts to address Arctic issues. For example, according to the 2013 *National Arctic Strategy Implementation Plan*, the United States will lead the development of the 2015 ministerial declaration that will include recommendations related to priorities for the U.S. chairmanship.

CONCLUSION

The Arctic region's rapid changes alter the environment, provide economic opportunities, and affect the people living there. With these changes and an increase in global attention on the region, the Council's work will likely continue to expand. The federal government has increased its focus on Arctic issues, and U.S. participation in the Council provides a forum to address some of these issues, collaborate domestically and internationally, and leverage resources. State leads U.S. participation in the Council and relies heavily on the work of other federal agencies for most of the U.S. contributions to the Council. However, State has not established a joint strategy for U.S. participation that outlines a clear direction or specifies necessary resources. Without a joint strategy, key collaborating agencies may continue to face challenges prioritizing the work, delivering unified messages, and consistently participating in the Council. In addition, the U.S. chairmanship will span two presidential administrations, making it particularly important to plan for continuity.

Furthermore, as Arctic issues have emerged and become a priority for the United States, State has increased its attention on the voluntary recommendations the United States has adopted through the Council and identified a need to more formally follow up on and assess progress toward implementing them. Because State does not have a process to review and track progress made by U.S. agencies to implement recommendations, State faces challenges planning for and prioritizing future areas of action on Arctic issues within the Council's purview. This will become increasingly important when the United States chairs the Council starting in 2015, providing State with valuable opportunities to shape future Council priorities.

Finally, State has recognized the importance of strengthening the Council's ability to make progress toward addressing Arctic issues through implementing recommendations and identified the need to more clearly specify actions with measurable outcomes and prioritize recommended actions. Specifically, State officials said the Council does not have guidelines for adopting clear and prioritized recommendations. Having no such guidelines has led to the Council adopting broad and numerous recommendations that the Arctic States face challenges implementing. As a result, State officials said that the Council has produced valuable work, but the recommendations have not historically produced actions leading to measurable results in addressing Arctic issues.

Recommendations for Executive Action

To help clarify the direction of future U.S. participation and position the United States for a successful Arctic Council chairmanship, assess the status of recommendations adopted through the Council, and strengthen the Council's ability to address Arctic issues within its purview, we recommend that the Secretary of State take the following three actions:

- As a part of its responsibilities in assuming the Council chair in 2015 and in collaboration with other relevant agencies, develop a joint strategy for U.S. participation in the Council that outlines a clear direction for the agencies and identifies resources needed to sustain collaborative efforts and consistent participation in the Council.
- Develop a process to review and track U.S. progress in implementing existing and any future recommendations.
- Work with other Arctic States to develop guidelines for producing clear recommendations with measurable actions and prioritizing the recommendations.

Agency Comments

We provided EPA, Department of Commerce, Department of Energy, Department of Homeland Security, the Department of the Interior, State, and USGCRP with a draft of this report for review and comment. In written comments. State generally agreed with our recommendations. State also noted that the Council recommendations we discuss in the report are expressly voluntary and should be differentiated from more formal commitments made through the international agreements. We modified the text to make this distinction.

EPA, Department of Commerce, Department of Energy, Department of Homeland Security, the Department of the Interior, and USGCRP provided technical comments that we incorporated as appropriate.

J. Alfredo Gómez
Director,
Natural Resources and Environment

Appendix I: Objectives, Scope, and Methodology

We were asked to review matters related to U.S. participation in the Arctic Council (Council). This report examines (1) how the Council is organized and how it addresses environmental and economic development issues; (2) how U.S. federal agencies with a key leadership role participate in the Council and the associated challenges, if any; and (3) the actions the Department of State (State) and other federal agencies have taken to implement and manage voluntary recommendations the United States has adopted through the Council and associated challenges, if any.

To choose from the many federal agencies involved in Council work and to scope our work, we selected agencies that hold key leadership roles. We chose State because it leads U.S. participation in the Council and agencies that head the U.S. delegations to the six Council working groups. These agencies are the U.S. Global Change Research Program, the Department of the Interior's U.S. Fish and Wildlife Service, the Department of Commerce's National Oceanic and Atmospheric Administration, the Environmental Protection Agency, and the Department of Energy's National Nuclear Security Administration. To provide information to answer all three of our objectives, we interviewed the following:

1) knowledgeable officials from the key federal agencies with a leadership role in the Council;
2) leaders of Permanent Participant organizations that represent Alaska Natives in the Council, including the Aleut International Association, Arctic Athabaskan Council, Gwich'in Council International, and Inuit Circumpolar Council; and
3) key Arctic stakeholders.

To select key Arctic stakeholders, we first identified a potential list of over 100 contacts by conducting research, as well as collecting referrals through a snowballed process, starting with key agency officials and stakeholders known to have direct involvement in the Council. Our list included contacts who represented diverse perspectives such as federal agencies, Alaska state agencies, nongovernmental organizations, groups representing Alaska Natives, science and policy research organizations, industry groups, and academic institutions. We ranked each contact from highest to lowest priority based on whether they: (1) had significant involvement in the Council – the six key agencies, Alaska-based Permanent Participants, and the state of Alaska; (2) were referred to us based on their experience working with the Council in the past; (3) worked with the Council as identified by Council documents or individual biographical research, but were not referred to us; and (4) represented an organization that applied for Council Observer status, or was in the process of doing so during our review. Once we organized our stakeholder list based on these rankings, we categorized the list by topic areas to acquire a balance of perspectives. These topic areas included environmental groups, federal agencies, health and human development groups, indigenous groups, industry, local or state agencies, policy groups, and science and research groups. Finally, we narrowed our list of stakeholders to those who were the highest priority based on our rankings, could offer a diverse perspectives, and had availability while ensuring alignment with GAO's allotted staff resources and time. Because of our scope and resources, we did not interview all stakeholders. Findings from our report based on stakeholder perspectives are not generalizable.

In addition, to identify how the Council is organized and addresses environmental and economic development issues, we attended a Council Senior Arctic Officials meeting in Stockholm, Sweden, in March 2013, and we analyzed Council documents, academic publications, and the Council's website. We also reviewed Arctic State publications, peer-reviewed academic literature, and nongovernmental organization reports published from 2008 to 2013. We determined these sources were sufficiently sound for our purpose. Furthermore, we focused on Arctic environmental and economic issues because the Council's mission outlines these as issues to address. We did not assess the effectiveness of the Council.

In addition to conducting interviews with key U.S. federal agency officials to determine how they participate in the Council and identify any associated challenges, we reviewed relevant agency and Council documents. We also conducted site visits to the Arctic Alaskan communities of Barrow, Kivalina, Kotzebue, and Wainwright. We met with community leaders to learn their perspectives on environmental and economic issues facing their community and how, if at all, key federal agencies considered their interests in their Council work. We selected these four communities based on the following factors: (1) stakeholder referrals, (2) location above the Arctic Circle, (3) representation of a range of community sizes and perspectives, and (4) accessibility in order to accommodate GAO's allotted staff and financial resources. Findings from these site visits cannot be generalized to those we did not include in our nonprobability sample. In addition to these four locations, we visited Anchorage, Alaska, where we interviewed knowledgeable Alaska state officials, key Arctic stakeholders, and two Permanent Participant group representatives to help inform our objectives. Finally, to identify and assess the actions that State and other federal agencies take to implement and manage voluntary recommendations the United States has adopted through the Council, we first analyzed Council ministerial declarations from 1998 to 2013 to develop a list of statements that directed actions to Arctic States. To create the list, two analysts independently identified these statements, compared results, and resolved any differences. We then asked State officials to review our list and identify which statements they viewed as Council recommendations to Arctic States. We also analyzed Council reports endorsed in the recommendations and the two multilateral agreements negotiated through the Council. Finally, we selected examples of actions agencies took to implement Council recommendations and multilateral agreements. This report will contribute to a multilateral audit of the Council led by the Supreme Audit Institutions of Norway and Russia, also with participation by Denmark, and Sweden.[1] The Supreme Audit Institutions of Canada, Finland, and Iceland observed the multilateral audit.

We conducted this performance audit from November 2012 to May 2014 in accordance with generally accepted government auditing standards. Those standards require that we plan and perform the audit to obtain sufficient, appropriate evidence to provide a reasonable basis for our findings and conclusions based on our audit objectives. We believe that the evidence obtained provides a reasonable basis for our findings and conclusions based on our audit objectives.

APPENDIX II: ARCTIC COUNCIL CHAIRMANSHIPS

Arctic state	Chairmanship years	Location of ministerial meeting
Canada	1996-1998	Iqaluit, Canada
United States of America	1998-2000	Barrow, Alaska
Finland	2000-2002	Inari, Finland
Iceland	2002-2004	Reykjavik, Iceland
Russia	2004-2006	Salekhard, Russia
Norway	2006-2009	Tromsø, Norway
Denmark	2009-2011	Nuuk, Greenland
Sweden	2011-2013	Kiruna, Sweden
Canada	2013-2015	To be determined

Source: Arctic Council.

Appendix III: Changes in September Ice Coverage from 2001 to 2013, Compared with the 1979-2000 Median September Ice Coverage

Figure 7 shows changes in September sea ice coverage, by year from 2001 through 2013.

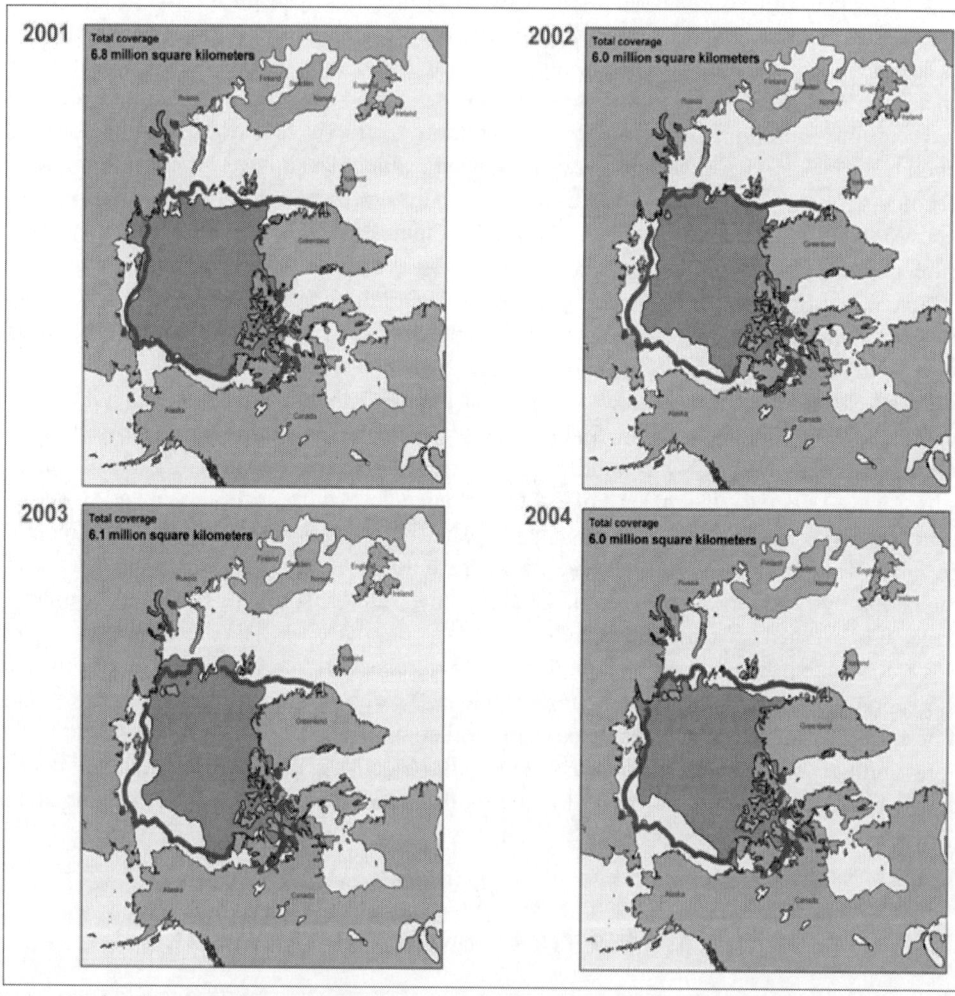

Source: National Snow and Ice Data Center.

Figure 7. September Sea Ice Coverage, 2001 through 2013.

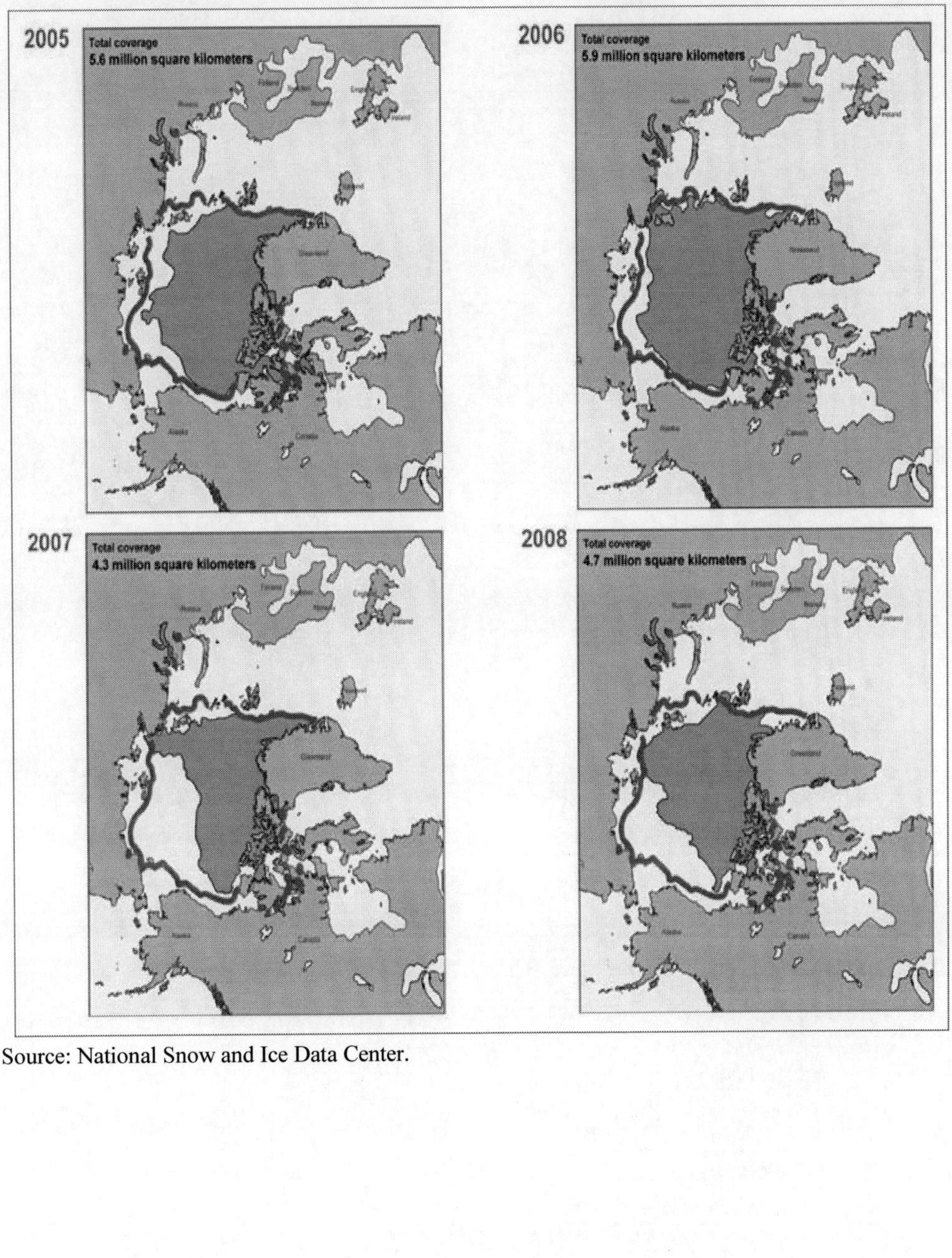

Source: National Snow and Ice Data Center.

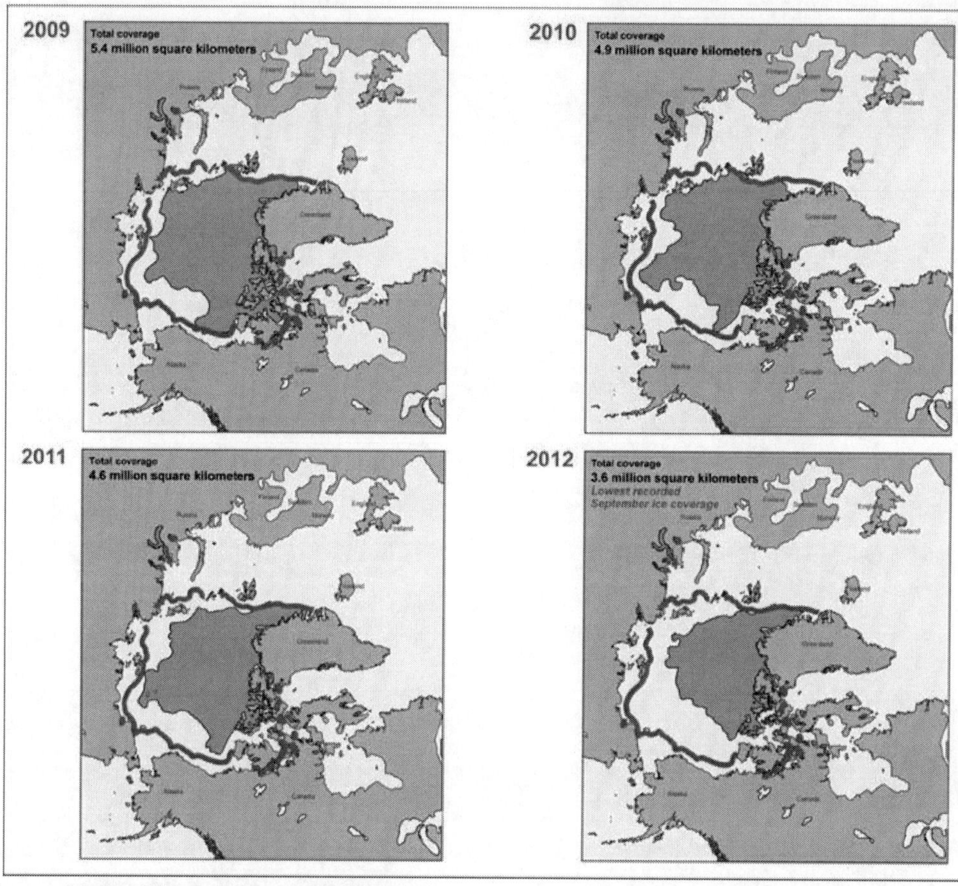

Source: National Snow and Ice Data Center.

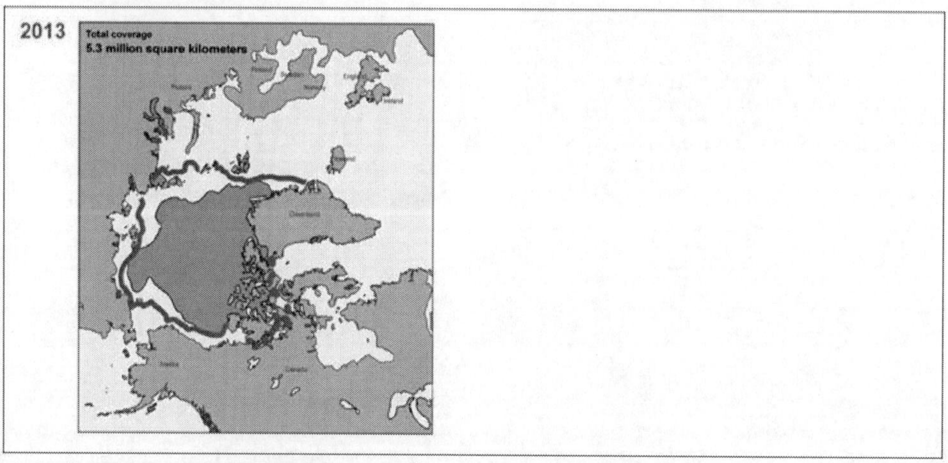

Source: National Snow and Ice Data Center.

APPENDIX IV: U.S. AGENCIES PARTICIPATING IN ARCTIC COUNCIL WORKING GROUPS AND TASK FORCES AS OF JANUARY 2014

Arctic Council working groups	
Arctic Contaminants Action Program	
	Department of Energy, Office of Energy Efficiency and Renewable Energy
	Department of State
	Environmental Protection Agency
Arctic Monitoring and Assessment Program	
	Department of Commerce, National Oceanic and Atmospheric Administration
	Department of Energy
	Department of Health and Human, Services Centers for Disease Control and Prevention
	Department of the Interior, U.S. Geological Survey
	Environmental Protection Agency
	Executive Office of the President, Office of Science and Technology Policy
	Executive Office of the President, U.S. Global Change Research Program
	National Aeronautics and Space Administration
	National Science Foundation
	United States Department of Agriculture, U.S. Forest Service
Conservation of Arctic Flora and Fauna	
	Department of Commerce, National Oceanic and Atmospheric Administration
	Department of Homeland Security, U.S. Coast Guard
	Department of the Interior, Bureau of Land Management/ North Slope Science Initiative
	Department of the Interior, U.S. Fish and Wildlife Service
	Executive Office of the President, Office of Science and Technology Policy
	Marine Mammal Commission
	National Aeronautics and Space Administration
	National Science Foundation
Emergency Prevention, Preparedness and Response	
	Department of Energy, National Nuclear Security Administration
	Department of Health and Human Services, Centers for Disease Control andPrevention
	Department of Homeland Security, U.S. Coast Guard
	Department of the Interior, Bureau of Ocean Energy Management
	Department of the Interior, Bureau of Safety and Environmental Enforcement
	Environmental Protection Agency
Protection of the Arctic Marine Environment	
	Department of Commerce, National Oceanic and Atmospheric Administration

Appendix IV. (Continued)

	Department of Defense, U.S. Navy
	Department of Homeland Security, U.S. Coast Guard
	Department of the Interior, Bureau of Safety and Environmental Enforcement
	Department of the Interior, Bureau of Ocean Energy Management
	Department of the Interior, U.S. Fish and Wildlife Service
	Department of State
	Environmental Protection Agency
	Marine Mammal Commission
Sustainable Development Working Group	
	Department of Commerce, National Oceanic and Atmospheric Administration
	Department of Energy
	Department of Health and Human Services, Centers for Disease Control and Prevention
	Department of Health and Human Services, Office of Global Affairs
	Department of Health and Human Services, Substance Abuse and Mental HealthServices Administration
	Department of the Interior, Bureau of Indian Affairs
	Department of the Interior, Bureau of Ocean Energy Management
	Department of the Interior U.S. Geological Survey
	Department of State
	Environmental Protection Agency
	National Science Foundation
	U.S. Arctic Research Commission
Arctic Council task forces	
Task Force on Black Carbon and Methane	
	Environmental Protection Agency
	Department of State
Task Force on Science Cooperation	
	Department of Commerce, National Oceanic and Atmospheric Administration
	Department of Defense, U.S. Navy, Office of Naval Research
	Department of Energy
	Department of the Interior, Bureau of Ocean Energy Management
	Department of the Interior, Fish and Wildlife Service
	Department of the Interior, U.S. Geological Survey
	Department of State
	Environmental Protection Agency
	Executive Office of the President, Office of Science and Technology Policy
	National Aeronautics and Space Administration
	National Science Foundation
	U.S. Arctic Research Commission

Task Force on Arctic Marine Oil Pollution Prevention	
	Department of Commerce, National Oceanic and Atmospheric Administration
	Department of Energy
	Department of Homeland Security, U.S. Coast Guard
	Department of the Interior, Bureau of Ocean Energy Management
	Department of the Interior, Bureau of Safety and Environmental Enforcement
	Department of State
	Environmental Protection Agency
Task Force to Create a Circumpolar Business Forum	
	Department of Homeland Security, U.S. Coast Guard
	Department of State
	U.S. Arctic Research Commission

Source: Department of State

APPENDIX V: ARCTIC POLICY GROUP PARTICIPANTS AS OF DECEMBER 2013

Executive Office of the President	
	National Ocean Council
	National Security Council
	Office of Science and Technology Policy
	U.S. Global Change Research Program
Federal agencies	
Department of Commerce	National Oceanic and Atmospheric Administration
Department of Defense	Joint Base Elmendorf-Richardson
	Joint Chiefs of Staff
	Office of the Secretary of Defense
	U.S. Air Force
	U.S. European Command
	U.S. Navy
	U.S. Northern Command
Department of Energy	National Nuclear Security Administration
	Office of International Affairs
	Office of Science
	National Nuclear Security Administration
Department of Health and Human Services	Centers for Disease Control and Prevention
	National Institutes of Health
	Substance Abuse and Mental Health Services Administration
Department of Homeland Security	Federal Emergency Management Agency
	U.S. Coast Guard

Appendix V. (Continued)

Department of the Interior	Bureau of Indian Affairs
	Bureau of Land Management - North Slope Science Initiative
	Bureau of Ocean Energy Management
	Bureau of Safety and Environmental Enforcement
	National Park Service
	Office of the Secretary
	U.S. Fish and Wildlife Service
	U.S. Geological Survey
Department of Justice	——
Department of State	——
Department of Transportation	Federal Aviation Administration
	Maritime Administration
Other federal entities	
	Environmental Protection Agency
	Marine Mammal Commission
	National Aeronautics and Space Administration
	National Science Foundation
	Office of the Federal Coordinator for Alaska Natural Gas Transportation Projects
	Smithsonian Institution
	U.S. Arctic Research Commission
State of Alaska	
	Office of the Governor
	Office of the Lieutenant Governor

APPENDIX VI: FULL TEXT OF FIGURE 6, ARCTIC ALASKA SITE VISITS

Barrow

Name: Barrow (traditionally known as Ukpeagvik)

Population: 4,212, of which approximately 61 percent are American Indian and Alaska Native, which includes Iñupiat Eskimos.

Location: Located on the Chukchi Sea coast, Barrow is the northernmost community in the United States.

Brief background: During the 1940s and 1950s, the military played an influential role in the Barrow area and constructed a Distant Early Warning line – a system of radar stations established to detect incoming Soviet bombers during the cold war and provide early warning of incursions into North American airspace. Exploration in the National Petroleum Reserve, which is a 22.8-million acre area with oil on Alaska's North Slope, brought new people to the region, and construction of the Prudhoe Bay oilfields and Trans-Alaska Pipeline have each contributed to its development. Since that time, tax revenues from the North Slope oil fields fund services in the area. Barrow has become the economic, transportation, and administrative center for Alaska's North Slope region.

Main industries providing employment: According to 2008 to 2012 data, out of 1,969 employed, 23 percent were in educational services, health care, and social assistance; 22 percent were in public administration; and 13 percent were in retail trade.

Alaska Native concerns on various Arctic issues: Some community leaders said they were concerned about the affects of increased Arctic activities and change in the climate on their access to subsistence resources, as well as limited plans and regulations to guide oil exploration and response to spills.

Kivalina

Name: Kivalina (traditionally known as Kivaliniq)

Population: 374, of which approximately 96 percent are American Indian and Alaska Native, which includes Iñupiat Eskimos.

Location: Kivalina is at the tip of an 8-mile barrier reef located between the Chukchi Sea and Kivalina River. It is 80 air miles northwest of Kotzebue and approximately 50 miles from the Red Dog Mine.

Brief background: The first recorded history of Kivalina occurred in 1847 when a Russian naval officer mistook a seasonal hunting camp at the north end of Kivalina Lagoon as a permanent settlement. Kivalina was located at its current location in 1905 when the Bureau of Indian Affairs built a school on the southern tip of the island. This compelled people who lived in the original Kivalina to migrate to the Kivalina created by the Bureau of Indian Affairs.

Main industries providing employment: According to 2008 to 2012 data, out of 132 employed, 24 percent were in educational services, health care, and social assistance; 23 percent were in retail trade; and 17 percent were in public administration.

Alaska Native concerns on various Arctic issues: Community leaders said severe coastal erosion has significantly affected Kivalina. The village intends to relocate to a new site 2.5 miles away. In addition, homes do not have full plumbing, and there is only one public space with three showers available.

Kotzebue

Name: Kotzebue (traditionally known as Kikiktagruk)

Population: 3,201, of which approximately 74 percent are American Indian and Alaska Native, which includes Iñupiat Eskimos.

Location: Kotzebue lies on the Baldwin Peninsula in Kotzebue Sound where the Noatak, Kobuk, and Selawik rivers end.

Brief background: The Iñupiat have occupied the area for at least 600 years. The site was a trading location for local natives for hundreds of years. Expansion of economic activities, such as oil and mineral exploration, and services to support these activities contributed to Kotzebue's development. Kotzebue is the service and transportation center for all villages in Alaska's northwest region.

Main industries providing employment: According to 2008 to 2012 data, out of 1,326 employed, 38 percent were in educational services, health care and social assistance; 13 percent were in transportation, warehousing, and utilities; and 11 percent were in public administration.

Alaska Native concerns on various Arctic issues: Community leaders said they have seen changes in sea ice, which affects the community's traditional subsistence lifestyle. As a result, many have turned to participating in the cash based economy, although there is strong desire to return to their traditional lifestyle.

Wainwright

Name: Wainwright (traditionally known as Olgoonik)

Population: 556, of which approximately 92 percent are American Indian and Alaska Native, which includes Iñupiat Eskimos.

Location: Wainwright sits on a wave-eroded coastal bluff of a narrow peninsula, which separates Wainwright Inlet from the Chukchi Sea. Wainwright is about 70 miles southwest of Barrow.

Brief background: The region around Wainwright was traditionally well populated, though the present village was not established until 1904 – when the Alaska Native Service built a school and instituted medical and other services. Wainwright is the third largest village in Alaska's North Slope region.

Main industries providing employment: According to 2008 to 2012 data, out of 233 employed, 25 percent were in educational services, health care, and social assistance; 15 percent were in construction; and 13 percent were in transportation, warehousing, and utilities.

Alaska Native concerns on various Arctic issues: Community leaders said they were concerned about the effects of climate change in their community and were in the process of constructing a seawall to protect a bluff from wave erosion. Leaders also said they would like to increase responsible oil and gas exploration to help support the community's lifestyle.

Appendix VII: Arctic Council Declaration Statements Identified by Department of State Officials as Recommendations to Arctic States

Year	Declaration	No.	Declaration statement
1998	Iqaluit	1.	Reaffirm our commitment from the Alta Declaration to take the findings and recommendations from the Arctic Monitoring and Assessment Programme Report, Arctic Pollution Issues: A State of the Arctic

Year	Declaration	No.	Declaration statement
			Environment Report, into consideration in our policies and programmes, to increase our effortsto limit and reduce emissions of contaminants into the environment and to promote international cooperation and make a determined effort to secure support for international actions in order to address the serious pollution risks reported by the Arctic Monitoring and Assessment Programme.
2000	Barrow	2.	Endorse and adopt the Arctic Council Plan to Eliminate Pollution in the Arctic.
		3.	Encourage actions to reduce the risks of radioactivity to the Arctic.
2002	Inari	4.	Acknowledge the recommendations arising from the Conservation of Arctic Flora and Fauna report Arctic Flora and Fauna, Status and Conservation as a strategy for future biodiversity conservation work of the Arctic Council.
		5.	The Ministers took note of recommendations generated by projects on timberline forests, sustainablereindeer husbandry, and sacred sites and encourage further dialogue among stakeholders on this basis.
		6.	Welcome with appreciation Arctic Monitoring and Assessment Programme's "Arctic Pollution 2002" Report and take the recommendations into consideration in our policies and programmes to reduce pollution affecting the Arctic.
2004	Reykjavik	7.	Encourage Member States to take effective measures to adapt to and manage the environmental, economic and social impacts of climate change and ultraviolet radiation, inter alia through enhancing the access of Arctic residents to information, decision makers and institutional capacity building.
		8.	Encourage relevant national and international research bodies and sponsors to take into account the Arctic Climate Impact Assessment science recommendations in the planning, development, and implementation of their programmes.
2004	Reykjavik	9.	Endorse the Arctic Marine Strategic Plan and encourage its implementation through the working groups and other mechanisms and in cooperation with regional and global bodies.
		10.	Encourage Member States and others to disseminate the Arctic Waters Oil Transfer Guidelines widely.
		11.	Direct Member States and the relevant working groups of the Arctic Council to consider appropriate follow-up actions [to the Arctic Human Development Report].

Appendix VII. (Continued)

Year	Declaration	No.	Declaration statement
2006	Salekhard	12.	Reconfirm [Arctic States'] commitments to the Reykjavik Declaration and to the Arctic Climate ImpactAssessment policy document, adopted at the Arctic Council meeting in 2004, and that the Member States will continue their active efforts to implement the recommendations on mitigation, adaptation, research, monitoring and outreach.
		13.	Urge Member States and other entities to strengthen monitoring and research efforts needed to comprehensively address Arctic change and to promote the establishment of a circumpolar Arctic observing network of monitoring stations with coordinated data handling and information exchange for scientific data, statistics, and traditional knowledge as a lasting legacy of the International Polar Year (and as the evolving Arctic component of the Global Earth Observing System of Systems).
		14.	Encourage research and practical actions, as well as exchange of expertise and best practices among the Member States to increase stability of the Arctic infrastructure in changing climatic conditions.
		15.	Urge all the Member countries to maintain and extend long-term monitoring of change in all parts of the Arctic.
2006	Salekhard	16.	Encourage the Arctic Council Member States to implement the Assessment Report on Acidifying Pollutants, Arctic Haze, and Acidification in the Arctic recommendations addressing acidification and Arctic haze effects, as appropriate, recognizing that the effects are regional in nature.
		17.	Endorse long-term monitoring of Arctic biodiversity to provide policymakers with the information needed to accurately assess the impacts from global environmental change and increased human activities related to regional development and economic growth.
		18.	Encourage countries to contribute actively to Circumpolar Biodiversity Monitoring Program and expect the program to provide valuable data for increased knowledge and improved management of biodiversity in the Arctic.
		19.	Encourage the Arctic Council Member States to focus on the following areas and initiate new projects in order to improve the capacity to respond to emergencies in the Arctic: • exchange of information, training and experience • public information• technical development and support, and • coordination of response

2009	Tromsø	20.	Urge implementation of early actions where possible on methane and other short-lived climate forcers
		21.	Urge the Arctic Council members to strengthen their work on adaptation to climate change, including by pursuing community-level actions and continue to share information on best practices
		22.	Encourage collaboration with the Methane to Markets Partnership and other relevant international bodies taking action to reduce methane and other short-lived forcers
		23.	Encourage the continuation of this work [the Sustaining Arctic Observing Networks] with emphasis on improving sustained long-term observation
		24.	Approve the Arctic Marine Shipping Assessment 2009 Report including its recommendations on enhancing Arctic marine safety, protecting Arctic people and environment and building Arctic marine infrastructure
		25.	Urge all States to apply the Arctic Council Offshore Oil and Gas Guidelines throughout the Arctic as minimum standards in national regulations
		26.	Encourage the continued implementation of the Arctic Council Programme of Action for the Protectionof the Arctic Marine Environment from Land-based Activities
		27.	Approve the findings and recommendations of the assessment of Oil and Gas Activities in the Arctic: Effects and Potential Effects
		28.	Decide to strengthen cooperation on prevention of, and response to, accidental spills of oil and hazardous substances in the Arctic
		29.	Urge Member States to apply the precautionary approach and polluter-pays principle as reflected in Principles 15 and 16 of the Rio Declaration, respectively, and conduct risk and environmental impact assessments for the exploration, development, transport and storage of oil, and enact and/or enforce appropriate laws and controls
2011	Nuuk	30.	Encourage Arctic States to implement, as appropriate in their national circumstances, relevant recommendations [from the Arctic Council Report on Short-Lived Climate Forcers] for reducing emissions of black carbon
2013	Kiruna	31.	Welcome the report on short-lived climate forcers and support its recommendations including that national black carbon emission inventories for the Arctic should continue to be developed and reported as a matter of priority
		32.	Announce the Agreement on Cooperation on Marine Oil Pollution Preparedness and Response in the Arctic and encourage future national, bi-national, and multinational contingency plans, training and exercises, to develop effective response measures

Appendix VII. (Continued)

Year	Declaration	No.	Declaration statement
2013	Kiruna	33.	Welcome the Recommended Practices in Prevention of Arctic Marine Oil Pollution Project reports and recommendations to Ministers, and encourage Arctic States to pursue further work in the recommended areas
		34.	Welcome the Arctic Biodiversity Assessment, the first Arctic-wide comprehensive assessment of status and emerging trends in Arctic biodiversity, approve its recommendations, and encourage ArcticStates to follow up on its recommendations
		35.	Encourage Arctic States to take decisive action to help sustain Arctic biodiversity and implement internationally agreed biodiversity objectives, to cooperate on adaptive management strategies for vulnerable species and ecosystems, and to continue existing Arctic biodiversity research and monitoring efforts through the Circumpolar Biodiversity Monitoring Program
		36.	Welcome the Arctic Ocean Acidification assessment, approve its recommendation and request the Arctic States to continue to take action on mitigation and adaptation and to monitor and assess the state of the Arctic Ocean acidification
		37.	Welcome the Arctic Ocean Review report, undertaken to provide guidance to Arctic States on strengthening governance in the Arctic through a cooperative, coordinated, and integrated approach to the management of the Arctic marine environment, approve its recommendations and request appropriate follow-up action, and report on progress at subsequent ministerial meetings
		38.	Encourage Arctic States to continue monitoring and assessment activities and enhance their efforts to meet the objectives of the Stockholm Convention
		39.	Welcome the report on Ecosystem Based Management, approve the definition, principles and recommendations, encourage Arctic States to implement recommendations both within and across boundaries

Source: GAO analysis of Department of State information

End Notes

[1] In general, the Arctic is the polar region located at the northernmost part of the Earth. Arctic stakeholders define the Arctic geographical area in different ways. For example, the Arctic Research and Policy Act of 1984 defines the Arctic as all U.S. and foreign territory north of the Arctic Circle and all U.S. territory north and west of the boundary formed by the Porcupine, Yukon, and Kuskokwim Rivers [in Alaska]; all contiguous

seas, including the Arctic Ocean and the Beaufort, Bering, and Chukchi Seas; and the Aleutian Chain. Pub. L. No. 98-373, 98 Stat. 1242,1248 (1984), *codified at* 15 U.S.C. § 4111. The Arctic Circle is the line of latitude located at 66° 33' 44" North of the equator. Other definitions of the Arctic use markers such as the southernmost extent of winter sea ice for oceanic boundaries, or the northern-most tree line for terrestrial boundaries.

[2] Thomas R. Karl, Jerry M. Melillo, and Thomas C. Peterson, eds., *Global Climate Change Impacts in the United States* (Cambridge University Press: 2009).

[3]The National Oceanic and Atmospheric Administration established the National Snow and Ice Data Center in 1976 to support research into the cryosphere—places on earth where water is frozen into ice or snow. For more information, see http://www.nsidc.org.

[4] U.S. Geological Survey Fact Sheet 2008-3049: Circum-Arctic Resource Appraisal: Estimates of Undiscovered Oil and Gas North of the Arctic Circle. http://pubs.usgs.gov/fs/2008/3049/.

[5] A 2013 interagency report on Arctic Management states that a subsistence way of life relies upon natural resources for food, shelter, clothing, transportation, and the maintenance of cultural traditions. J.P. Clement, J.L. Bengston, and B.P. Kelly, Interagency Working Group on Coordination of Domestic Energy Development and Permitting in Alaska (D.J. Hayes, Chair), *Managing for the Future in a Rapidly Changing Arctic: A report to the President* (Washington, D.C.: March 2013).

[6] The Arctic States all have territory north of the Arctic Circle. The Kingdom of Denmark includes Greenland and the Faroe Islands.

[7] Through the Arctic Environmental Protection Strategy, the Arctic States agreed to, among other things, cooperate in scientific research, assess potential environmental impacts of development activities, and to hold regular meetings to assess the progress made and to coordinate future action.

[8] The White House, *2009 National Security Presidential Directive-66/Homeland Security Presidential Directive-25* (NSPD-66/HSPD-25), (Washington, D.C.: Jan. 9, 2009); the White House, *National Strategy for the Arctic Region* (Washington, D.C.: May 10, 2013).

[9] These six agencies with a leadership role head the U.S. delegations to the Council's ongoing work.

[10] GAO, *Alaska Native Villages: Limited Progress Has Been Made on Relocating Villages Threatened by Flooding and Erosion*, GAO-09-551 (Washington, D.C.: June 3, 2009).

[11] GAO, *Maritime Infrastructure: Key Issues Related to Commercial Activity in the U.S. Arctic over the Next Decade,* GAO-14-299 (Washington, D.C.: Mar. 19, 2014).

[12] The findings from our interviews with these communities cannot be generalized to those we did not include but provide examples of communities' interaction with federal agencies to address Arctic issues.

[13] The continental shelf is the seabed and subsoil of the submarine areas that extend either (1) beyond the state's territorial sea throughout the natural prolongation of its land territory to the outer edge of the continental margin or (2) to a distance of 200 nautical miles from the baselines from which the breadth of the territorial sea is measured where the outer edge of the continental margin does not extend up to that distance.

[14] White House, *National Strategy for the Arctic Region* (Washington, D.C.: May 10, 2013).

[15] Arctic Ocean Conference, *Ilulissat Declaration* (Ilulissat, Greenland: May 27-29, 2008).

[16] The Polar Code will cover the full range of design, construction, equipment, operational, training, search and rescue and environmental protection matters relevant to ships operating in the Arctic and Antarctic.

[17] White House, *Arctic Region Policy*, National Security Presidential Directive/NSPD-66 and Homeland Security Presidential Directive/HSPD-25 (Washington, D.C.: Jan. 9, 2009).

[18] White House, *National Strategy for the Arctic Region* (Washington, D.C.: May 10, 2013).

[19] White House, *Implementation Plan for the National Strategy for the Arctic Region* (Washington, D.C.: Jan. 30, 2014).

[20] J.P. Clement, J.L. Bengston, and B.P. Kelly, Interagency Working Group on Coordination of Domestic Energy Development and Permitting in Alaska (D.J. Hayes, Chair), *Managing for the Future in a Rapidly Changing Arctic. A report to the President* (Washington, D.C.: March 2013).

[21] C.J. Markon, S.F. Trainor, and F.S. Chapin III, eds., *The United States National Climate Assessment—Alaska Technical Chapter, U.S. Geological Survey Circular 1379* (2012).

[22] Thomas R. Karl, Jerry M. Melillo, and Thomas C. Peterson, eds., *Global Climate Change Impacts in the United States* (Cambridge University Press: 2009).

[23] Arctic Council Arctic Monitoring and Assessment Programme, *Arctic Climate Issues 2011: Changes in Arctic Snow, Water, Ice and Permafrost* (Oslo, Norway: 2012).

[24] Arctic Council Arctic Monitoring and Assessment Programme, *Arctic Pollution 2009* (Oslo: 2009); Arctic Council Arctic Monitoring and Assessment Programme, *AMAP Assessment 2011: Mercury in the Arctic*

(Oslo: 2011). Persistent organic pollutants and mercury are toxic chemicals that can accumulate in the environment and contaminate food sources.

[25] We have ongoing work related to effects of ocean acidification and related federal actions.

[26] For more information regarding Arctic shipping, refer to GAO-14-299.

[27] GAO-14-299.

[28] Arctic Council Task Force on Short-Lived Climate Forcers, *Progress Report and Recommendations for Ministers* (2011). Black carbon, which is soot, can warm the climate by darkening the snow and ice and absorbing sunlight, which leads to further warming and melting. Other short-lived climate forcers include ozone and methane, among other things.

[29] Throughout this report, we use the term indigenous to describe the communities of people that have historically resided in the Arctic. We use the term Alaska Natives to refer to the indigenous people who reside in Alaska.

[30] Specifically, the Council opens Permanent Participant status to other Arctic organizations of indigenous people with majority Arctic indigenous constituency that represent either an Arctic indigenous group in more than one Arctic state or represent multiple indigenous groups in one Arctic state. The declaration further states that the number of Permanent Participant groups at any time should be less than the number of Arctic Council member states.

[31] The six Permanent Participant Groups to the Arctic Council include: the Aleut International Association, Arctic Athabaskan Council, Gwich'in Council International, Inuit Circumpolar Council, Russian Arctic Indigenous Peoples of the North, and the Saami Council. Aleut International Association, Arctic Athabaskan Council, Gwich'in Council International, and Inuit Circumpolar Council represent Alaska Natives.

[32] For more information regarding the Arctic Council Indigenous Peoples' Secretariat, visit its website: http://www.arcticpeoples.org/.

[33] Arctic States permit Permanent Participants and Observers to attend the biennial ministerial meetings.

[34] Other ministers may serve as the Arctic Minister for an Arctic State. For example, Canada's Environment Minister leads the 2013-2015 Canadian chairmanship of the Arctic Council.

[35] In February 2014, State announced that it will name a special high-level representative to the Council tasked with advancing U.S. Arctic interests in preparation for the U.S. chairmanship of the Council beginning in May 2015.

[36] Arctic Council, *Impacts of a Warming Arctic: Arctic Climate Impact Assessment* (Cambridge, United Kingdom: October 15, 2004).

[37] Arctic Council, *Arctic Marine Shipping Assessment 2009 Report* (Tromsø, Norway: April, 2009).

[38] Victoria Gofman, *Community-Based Monitoring Handbook: Lessons from the Arctic and Beyond*, Conservation of Arctic Flora and Fauna, CBMP Report No.21 (Akureyri, Iceland: August, 2010).

[39] For additional information regarding the Arctic Contaminants Action Program's project on obsolete pesticides, see: http://www.arctic-council.org/index.php/en/acap-home/acappsgs/obsolete-pesticides.

[40] For additional information regarding the Arctic Maritime and Aviation Transportation Infrastructure Initiative, see: http://www.arcticinfrastructure.org.

[41] On March 26, 2014, the Arctic Council issued a statement announcing that SAOs agreed to move forward with the creation of the forum now referred to as the Arctic Economic Council. According to the statement, the Arctic Economic Council will be able to put forward proposals and reports to help inform the work of the Arctic Council.

[42] Prior to 2011, the United States Ministerial Representative was a lower-level State official. For example, the Deputy Secretary of State attended the 2009 ministerial meeting and the Under Secretary of State for Democracy and Global Affairs attended the 2006 ministerial meeting.

[43] The task force was established in the Tromsø Declaration at the Sixth Arctic Council's biennial ministerial meeting in April 2009 and, according to EPA officials, concluded its work in May 2013.

[44] For example, according to a NOAA official, the agency annually provides $20,000 to the Protection of the Arctic Marine Environment working group and FWS officials said the agency contributes $30,000 annually to the Conservation of Arctic Flora and Fauna working group.

[45] Under UNCLOS, coastal states have the right to regulate and authorize marine scientific research in their exclusive economic zone and on their continental shelf in accordance with the treaty. A coastal state's exclusive economic zone is the area beyond and adjacent to the territorial sea and extends no more than 200 nautical miles from the territorial sea baseline.

[46] In1971, President Nixon directed in National Security Decision Memorandum 144 that an Interagency Arctic Policy Group be established, chaired by State.

[47] According to an FWS official, the FWS delegation lead participated in the meetings by phone.

[48] GAO, *Results Oriented Government: Practices That Can Help Enhance and Sustain Collaboration Among Federal Agencies*, GAO-06-15 (Washington, D.C.: Oct. 21, 2005). Other key practices we identified that can enhance and sustain interagency collaboration are to define and articulate a common outcome; agree on roles and responsibilities; establish compatible policies, procedures, and other means to operate across agency boundaries; develop mechanisms to monitor, evaluate, and report on the results; reinforce agency accountability for collaborative efforts; and reinforce individual accountability for collaborative efforts. Additional discussion on interagency collaboration mechanisms can be found in GAO, *Managing for Results: Key Considerations for Implementing Interagency Collaborative Mechanisms*, GAO-12-1022 (Washington, D.C.: Sept. 27, 2012) and GAO, *Managing for Results: Implementation Approaches Used to Enhance Collaboration in Interagency Groups*, GAO-14-220 (Washington, D.C.: Feb.14, 2014).

[49] The White House, *National Strategy for the Arctic Region* (Washington, D.C.: May 10, 2013); J.P. Clement, J.L. Bengston, and B.P. Kelly, Interagency Working Group on Coordination of Domestic Energy Development and Permitting in Alaska (D.J. Hayes, Chair), *Managing for the Future in a Rapidly Changing Arctic: A report to the President* (Washington, D.C.: March 2013).

[50] In 2010, the Alaska State Legislature established the Alaska Northern Waters Task Force to identify opportunities to increase the state's engagement on issues associated with diminishing sea ice and associated economic and environmental affects. The task force included state legislators, leaders from Alaskan communities, and representatives of key state and federal agencies. The task force produced a report in January 2012, which included findings and recommendations.

[51] Alaska officials said additional priorities include preventing suicides and developing safe and sustainable sanitation facilities for smaller, isolated Arctic communities. In addition, in early 2014, both houses of the Alaska State Legislature approved a resolution urging State to consider Alaska's priorities when it chairs the Council and work in partnership with state officials to appoint a Council chair.

[52] According to the Institute of the North, a nongovernmental organization that aims to inform public policy and engage citizens, the Arctic Maritime and Aviation Transportation Infrastructure Initiative seeks to evaluate northern infrastructure ports, airports, and response capability by inventorying maritime and aviation assets in the Arctic. A multinational steering committee composed of government, academic, and private sector entities leads the initiative.

[53] At the time of our review, the Alaska Arctic Policy Commission was finalizing a report to outline the state of Alaska's overarching Arctic priorities. Alaska Arctic Policy Commission, *Preliminary Report to the Alaska State Legislature* (Anchorage and Bethel, Alaska: submitted Jan. 30, 2014).

[54] This working group addresses the environmental consequences and biological effects in the Arctic resulting from climate change, among other issues.

[55] For example, in 2012, an Aleut International Association representative said State provided $75,000 and, in 2013, it provided $55,000. Not all Permanent Participants use this funding. For example, Gwich'in Council International has not used this funding in recent years mainly because of limited number of staff, according to a Permanent Participant representative.

[56] We reported in 2009 that coastal erosion from rising temperatures and thawing permafrost has posed imminent threats to 31 Alaska villages, many of which have decided to relocate or explore relocation options (GAO-09-551).

[57] In addition to the recommendations made to the Arctic States, the Council makes recommendations to Council bodies, such as the working groups, and other stakeholders, such as the scientific community. This report focuses on the recommendations the Council makes to the Arctic States.

[58] Through the analysis, we identified 21 Ministerial Declaration statements that endorsed Arctic Council work products.

[59] Arctic Council, *Salekhard Declaration: On the Occasion of the Tenth Anniversary of the Arctic Council, the Fifth Arctic Council Ministerial Meeting* (Salekhard, Russia: Oct. 26, 2006).

[60] FWS leads the U.S. efforts to implement the Circumpolar Biodiversity Monitoring Program, with participation from NOAA and other agencies.

[61] Arctic Council, *Tromsø Declaration: On the Occasion of the Sixth Ministerial Meeting of the Arctic Council* (Tromsø, Norway: Apr. 29, 2009); Arctic Council Protection of the Arctic Marine Environment Working Group, *Arctic Offshore Oil and Gas Guidelines* (Akureyri, Iceland: Apr. 29, 2009).

[62] Arctic Council, *Nuuk Declaration: On the Occasion of the Seventh Ministerial Meeting of the Arctic Council* (Nuuk, Greenland: May 12, 2011); Arctic Council Secretariat, *Kiruna Declaration: On the Occasion of the Eighth Ministerial Meeting of the Arctic Council* (Kiruna, Sweden: May 15, 2013); Arctic Council Task Force on Short-Lived Climate Forcers, *An Assessment of Emissions and Mitigation Options for Black Carbon for the*

Arctic Council (April 2011); Arctic Council Task Force on Short-Lived Climate Forcers, *Recommendations to Reduce Black Carbon and Methane Emissions to Slow Arctic Climate Change* (May 2013).

[63] According to State officials, the United States signed the Arctic Search and Rescue Agreement as an executive agreement, which does not require Senate ratification to enter into force for the United States. The officials also said that the United States signed the Arctic Marine Oil Pollution Preparedness and Response Agreement but has not consented to be bound by it.

[64] *Agreement on Cooperation on Aeronautical and Maritime Search and Rescue in the Arctic* (Nuuk, Greenland: May 12, 2011).

[65] *Agreement on Cooperation on Marine Oil Pollution Preparedness and Response in the Arctic* (Kiruna, Sweden: May 15, 2013).

[66] Arctic Council, *Barrow Declaration: On the Occasion of the Second Ministerial Meeting of the Arctic Council* (Barrow, AK USA: Oct. 13, 2000); Arctic Council, *Tromsø Declaration* (2009).

[67] According to State officials, the Council has initiated some efforts under the Canadian chairmanship to begin inventorying and monitoring existing recommendations at the Council level. In addition, some Council working groups have development mechanisms to follow up on recommendations made in specific reports.

[68] Arctic Council, *Arctic Climate Impact Assessment Policy Document* (Reykjavik, Iceland: November, 2004); Arctic Council, *Regional Programme of Action for the Protection of the Arctic Marine Environment from Land-Based Activities* (Akureyri, Iceland: Apr. 29, 2009).

[69] According to EPA officials, the Arctic States have varying abilities to implement the Council's recommendations, so the Council includes the phrase "as appropriate in their national circumstances" in the language of some recommendations to Arctic States.

End Note for Appendix I

[1] A Supreme Audit Institution is the national audit office of a country. GAO is the Supreme Audit Institution of the United States.

In: The Arctic Council and a Changing Arctic
Editor: Hayley Justina Cobb
ISBN: 978-1-63321-719-5

Chapter 2

CHANGES IN THE ARCTIC: BACKGROUND AND ISSUES FOR CONGRESS*

Ronald O'Rourke

SUMMARY

The diminishment of Arctic sea ice has led to increased human activities in the Arctic, and has heightened interest in, and concerns about, the region's future. The United States, by virtue of Alaska, is an Arctic country and has substantial interests in the region.

On May 10, 2013, the Obama Administration released a national strategy document for the Arctic region. On January 30, 2014, the Obama Administration released an implementation plan for this strategy.

Record low extents of Arctic sea ice over the past decade have focused scientific and policy attention on links to global climate change and projected ice-free seasons in the Arctic within decades. These changes have potential consequences for weather in the United States, access to mineral and biological resources in the Arctic, the economies and cultures of peoples in the region, and national security.

The five Arctic coastal states—the United States, Canada, Russia, Norway, and Denmark (of which Greenland is a territory)—are in the process of preparing Arctic territorial claims for submission to the Commission on the Limits of the Continental Shelf.

The Russian claim to the enormous underwater Lomonosov Ridge, if accepted, would reportedly grant Russia nearly one-half of the Arctic area. There are also four other unresolved Arctic territorial disputes.

The diminishment of Arctic ice could lead in coming years to increased commercial shipping on two trans-Arctic sea routes—the Northern Sea Route and the Northwest Passage. Current international guidelines for ships operating in Arctic waters are being updated.

---

* This is an edited, reformatted and augmented version of a Congressional Research Service publication, No. R41153, dated June 5, 2014.

Changes to the Arctic brought about by warming temperatures will likely allow more exploration for oil, gas, and minerals. Warming that causes permafrost to melt could pose challenges to onshore exploration activities. Increased oil and gas exploration and tourism (cruise ships) in the Arctic increase the risk of pollution in the region. Cleaning up oil spills in ice-covered waters will be more difficult than in other areas, primarily because effective strategies have yet to be developed.

Large commercial fisheries exist in the Arctic. The United States is currently meeting with other countries regarding the management of Arctic fish stocks. Changes in the Arctic could affect threatened and endangered species. Under the Endangered Species Act, the polar bear was listed as threatened on May 15, 2008. Arctic climate change is also expected to affect the economies, health, and cultures of Arctic indigenous peoples.

Two of the Coast Guard's three polar icebreakers—*Polar Star* and *Polar Sea*—have exceeded their intended 30-year service lives, and *Polar Sea* is not operational. The possibility of increased sea traffic through Arctic waters raises an issue concerning Arctic search and rescue capabilities.

On May 12, 2011, representatives from the member states of the Arctic Council signed an agreement on cooperation on aeronautical and maritime search and rescue in the Arctic.

The Arctic has increasingly become a subject of discussion among political leaders of the nations in the region. Although there is significant international cooperation on Arctic issues, the Arctic is also increasingly being viewed by some observers as a potential emerging security issue. In varying degrees, the Arctic coastal states have indicated a willingness to establish and maintain a military presence in the high north. U.S. military forces, particularly the Navy and Coast Guard, have begun to pay more attention to the region.

On May 21, 2013, the Coast Guard released a strategy document for the Arctic.

Introduction

The diminishment of Arctic sea ice has led to increased human activities in the Arctic, and has heightened interest in, and concerns about, the region's future. Issues such as Arctic sovereignty claims; commercial shipping through the Arctic; Arctic oil, gas, and mineral exploration; endangered Arctic species; and increased military operations in the Arctic could cause the region in coming years to become an arena of international cooperation or competition.

The United States, by virtue of Alaska, is an Arctic country and has substantial political, economic, energy, environmental, and other interests in the region. Decisions that Congress, the executive branch, foreign governments, international organizations, and commercial firms make on Arctic-related issues could significantly affect these interests.

Background[1]

Definitions of the Arctic

There are multiple definitions of the Arctic that result in differing descriptions of the land and sea areas encompassed by the term.

Policy discussions of the Arctic can employ varying definitions of the region, and readers should bear in mind that the definition used in one discussion may differ from that used in another. This CRS report does not rely on any one definition.

Arctic Circle Definition and Resulting Arctic Countries

The most common and basic definition of the Arctic defines the region as the land and sea area north of the Arctic Circle (a circle of latitude at about 66.34° North). For surface locations within this zone, the sun is generally above the horizon for 24 continuous hours at least once per year (at the summer solstice) and below the horizon for 24 continuous hours at least once per year (at the winter solstice).

The Arctic Circle definition includes the northernmost third or so of Alaska, as well as the Chukchi Sea, which separates that part of Alaska from Russia, and U.S. territorial and Exclusive Economic Zone (EEZ) waters north of Alaska. It does not include the lower two-thirds or so of Alaska or the Bering Sea, which separates that lower part of the state from Russia.

Eight countries have territory north of the Arctic Circle: the United States (Alaska), Canada, Russia, Norway, Denmark (by virtue of Greenland, a member country of the Kingdom of Denmark), Finland, Sweden, and Iceland.[2] These eight countries are often referred to as the Arctic countries, and they are the member states of the Arctic Council, an intergovernmental forum established in 1996.[3]

A subset of the eight Arctic countries are the five countries that are considered Arctic coastal states: the United States, Canada, Russia, Norway, and Denmark (by virtue of Greenland).

Definition in Arctic Research and Policy Act (ARPA) of 1984

Section 112 of the Arctic Research and Policy Act (ARPA) of 1984 (Title I of P.L. 98-373 of July 31, 1984)[4] defines the Arctic as follows:

> As used in this title, the term "Arctic" means all United States and foreign territory north of the Arctic Circle and all United States territory north and west of the boundary formed by the Porcupine, Yukon, and Kuskokwim Rivers [in Alaska]; all contiguous seas, including the Arctic Ocean and the Beaufort, Bering, and Chukchi Seas; and the Aleutian chain.

This definition, which is codified at 15 U.S.C. 4111,[5] includes certain parts of Alaska below the Arctic Circle, including the Aleutian Islands and portions of central and western mainland Alaska, such as the Seward Peninsula and the Yukon Delta.

Figure 1 below shows the Arctic area of Alaska as defined by ARPA.

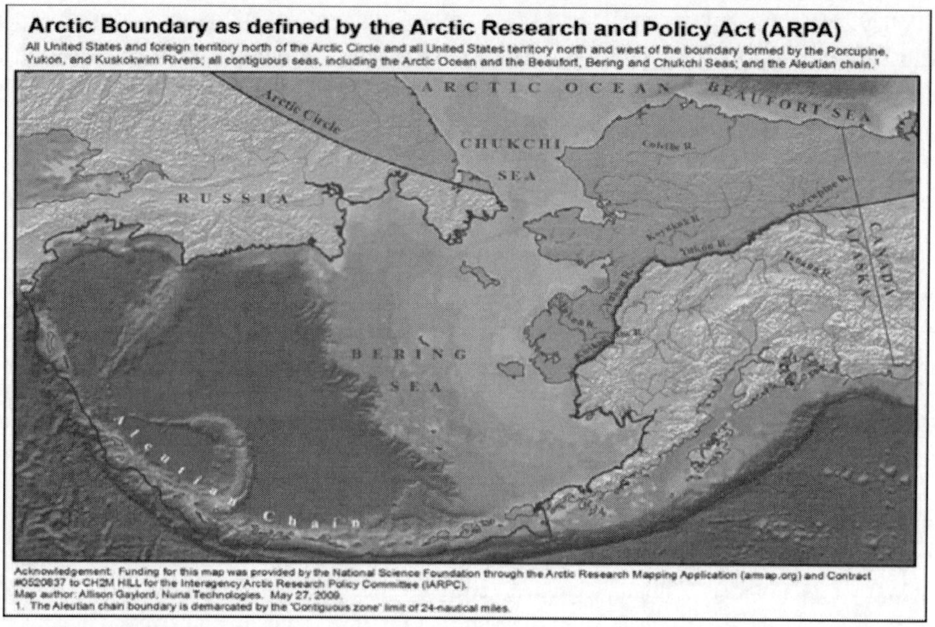

Source: U.S. Arctic Research Commission (http://www.arctic.gov/maps/ARPA_Alaska_only_150dpi.jpg, accessed on December 23, 2011).

Figure 1. Arctic Area of Alaska as Defined by ARPA.

Figure 2 shows the entire Arctic area as defined by ARPA.

Source: U.S. Arctic Research Commission (http://www.arctic.gov/maps/ARPA_Polar_150dpi.jpg, accessed on December 23, 2011).

Figure 2. Entire Arctic Area as Defined by ARPA.

Other Definitions

Other definitions of the Arctic are based on factors such as average temperature, the northern tree line, the extent of permafrost on land, the extent of sea ice on the ocean, or jurisdictional or administrative boundaries.[6] A definition based on a climate-related factor could circumscribe differing areas over time as a result of climate change.

The 10° C isotherm definition of the Arctic defines the region as the land and sea area in the northern hemisphere where the average temperature for the warmest month (July) is below 10° Celsius, or 50° Fahrenheit. This definition results in an irregularly shaped Arctic region that excludes some land and sea areas north of the Arctic Circle but includes some land and sea areas south of the Arctic Circle. This definition currently excludes all of Finland and Sweden, as well as some of Alaska above the Arctic Circle, while including virtually all of the Bering Sea and Alaska's Aleutian Islands.[7]

The definition of the Arctic adopted by the Arctic Monitoring and Assessment Programme (AMAP)—a working group of the Arctic Council—"essentially includes the terrestrial and marine areas north of the Arctic Circle (66°32' N), and north of 62° N in Asia and 60° N in North America, modified to include the marine areas north of the Aleutian chain, Hudson Bay, and parts of the North Atlantic, including the Labrador Sea."[8] The AMAP website includes a map showing the Arctic Circle, 10° C isotherm, tree line, and AMAP definitions of the Arctic.[9]

Some observers use the term "high north" as a way of referring to the Arctic. Some observers make a distinction between the "high Arctic"—meaning, in general, the colder portions of the Arctic that are closer to the North Pole—and other areas of the Arctic that are generally less cold and further away from the North Pole, which are sometimes described as the low Arctic or the subarctic.

U.S. Arctic Research

Arctic Research and Policy Act (ARPA) of 1984, As Amended

The Arctic Research and Policy Act (ARPA) of 1984 (Title I of P.L. 98-373 of July 31, 1984)[10] "provide[s] for a comprehensive national policy dealing with national research needs and objectives in the Arctic."[11] The act, among other things

- made a series of findings concerning the importance of the Arctic and Arctic research;
- established the U.S. Arctic Research Commission (USARC) to promote Arctic research and recommend Arctic research policy;
- designated the National Science Foundation (NSF) as the lead federal agency for implementing Arctic research policy;
- established the Interagency Arctic Research Policy Committee (IARPC) to develop a national Arctic research policy and a five-year plan to implement that policy, and designated the NSF representative on the IARPC as its chairperson;[12] and
- defined the term "Arctic" for purposes of the act.

The Arctic Research and Policy Act of 1984 was amended by P.L. 101-609 of November 16, 1990. For the texts of the Arctic Research and Policy Act of 1984 and P.L. 101-609, see **Appendix A** and **Appendix B**, respectively.

FY2015 NSF Budget Request for Arctic Research

NSF—the lead federal agency for implementing Arctic research policy (see "Arctic Research and Policy Act (ARPA) of 1984, As Amended")—carries out Arctic research activities through its Division of Polar Programs (PLR), which is part of its Directorate for Geosciences (GEO).[13] NSF is requesting a total of $435.11 million for PLR for FY2015 (an increase of 0.1% over the estimated FY2014 level), including $125.39 million for research in both the Arctic and Antarctic (an increase of 0.2% over the estimated FY2014 level) and $38.64 million for Arctic research support and logistics (a reduction of 5.4% from the estimated FY2014 level).[14]

NSF states in its FY2015 overview of GEO that

> As the primary U.S. supporter of fundamental research in the polar regions, GEO provides interagency leadership for U.S. polar activities. In the Arctic, NSF helps coordinate research planning as directed by the Arctic Research Policy Act of 1984. The NSF Director chairs the Interagency Arctic Research Policy Committee created for this purpose, which is now a component of the President's National Science and Technology Council (NSTC).[15]

NSF states in its FY2015 overview of PLR that

> The Division of Polar Programs (PLR) is the primary U.S. supporter of, and serves NSF interagency leadership responsibilities for, fundamental research in the polar regions. Arctic Sciences supports research in social, earth systems, and a broad range of natural sciences; its Research Support & Logistics program responds to research by assisting researchers with access to the Arctic and the planning and sharing of results with local Arctic communities....
>
> PLR's FY 2015 Request reflects three key priorities: (1) maintaining strong disciplinary programs that provide a base for investments in cross-disciplinary science programs; (2) focusing basic research on cross-foundation (e.g., SEES)[16] and interagency priorities; and (3) supporting and improving the efficiency of critical facilities that enable research in both polar regions.... For the Arctic, shared cross-directorate basic research objectives, the Interagency Arctic Research Policy Committee's (IARPC) *Arctic Research Plan: FY 2013-2017*,[17] and *the National Ocean Policy Implementation Strategy*[18] inform science investment priorities.
>
> As with most GEO divisions, PLR funds both research and the necessary research support in the form of logistics and infrastructure. The research budget is approximately 30 percent of the total division budget. Of this amount, 50 percent is available for new grants each year. The supporting logistics and infrastructure budget is 70 percent of the overall budget.[19]

Regarding its FY2015 request for $38.64 million for Arctic research support and logistics within PLR, NSF states:

> This program provides support for Arctic researchers, including airplanes, helicopters, access to icebreakers, and field camps for approximately 150 projects in

remote sites in Alaska, Canada, Arctic Scandinavia, Russia, and the Arctic Ocean. Summit Station on the Greenland ice cap operates as a year-round international site for a variety of atmospheric and geophysical measurements. Reduced funding (-$2.20 million, to a total of $38.64 million) will limit the use of marine platforms such as the newly available *RV SIKULIAQ*.[20]

Major U.S. Policy Documents Relating to the Arctic

January 2009 Arctic Policy Directive (NSPD 66/HSPD 25)

On January 12, 2009, the George W. Bush Administration released a presidential directive establishing a new U.S. policy for the Arctic region. The directive, dated January 9, 2009, was issued as National Security Presidential Directive 66/Homeland Security Presidential Directive 25 (NSPD 66/HSPD 25). The directive was the result of an interagency review, and it superseded for the Arctic (but not the Antarctic) a 1994 presidential directive on Arctic and Antarctic policy. The directive, among other things,

- states that the United States is an Arctic nation, with varied and compelling interests in the region;
- sets forth a six-element overall U.S. policy for the region;
- describes U.S. national security and homeland security interests in the Arctic; and
- discusses a number of issues as they relate to the Arctic, including international governance; the extended continental shelf and boundary issues; promotion of international scientific cooperation; maritime transportation; economic issues, including energy; and environmental protection and conservation of natural resources.

The Obama Administration has not issued a new directive superseding NSPD 66/HSPD 25; it is currently operating under the Bush Administration's policy directive.[21] For the text of NSPD 66/HSPD 25, see **Appendix C**.

May 2010 National Security Strategy

In May 2010, the Obama Administration released a national security strategy document that states:

> The United States is an Arctic Nation with broad and fundamental interests in the Arctic region, where we seek to meet our national security needs, protect the environment, responsibly manage resources, account for indigenous communities, support scientific research, and strengthen international cooperation on a wide range of issues.[22]

May 2013 National Strategy for Arctic Region

On May 10, 2013, the Obama Administration released a document entitled *National Strategy for the Arctic Region*.[23] The document appears to supplement rather than supersede the January 2009 Arctic policy directive (NSPD 66/HSPD 25) discussed above.[24] The executive summary of *National Strategy for the Arctic Region* begins by quoting the above statement from the May 2010 national security strategy document, and then states:

The *National Strategy for the Arctic Region* sets forth the United States Government's strategic priorities for the Arctic region. This strategy is intended to position the United States to respond effectively to challenges and emerging opportunities arising from significant increases in Arctic activity due to the diminishment of sea ice and the emergence of a new Arctic environment. It defines U.S. national security interests in the Arctic region and identifies prioritized lines of effort, building upon existing initiatives by Federal, state, local, and tribal authorities, the private sector, and international partners, and aims to focus efforts where opportunities exist and action is needed. It is designed to meet the reality of a changing Arctic environment, while we simultaneously pursue our global objective of combating the climatic changes that are driving these environmental conditions. Our strategy is built on three lines of effort:

1. **Advance United States Security Interests** – We will enable our vessels and aircraft to operate, consistent with international law, through, under, and over the airspace and waters of the Arctic, support lawful commerce, achieve a greater awareness of activity in the region, and intelligently evolve our Arctic infrastructure and capabilities, including ice-capable platforms as needed. U.S. security in the Arctic encompasses a broad spectrum of activities, ranging from those supporting safe commercial and scientific operations to national defense.

Pursue Responsible Arctic Region Stewardship – We will continue to protect the Arctic environment and conserve its resources; establish and institutionalize an integrated Arctic management framework; chart the Arctic region; and employ scientific research and traditional knowledge to increase understanding of the Arctic.

Strengthen International Cooperation – Working through bilateral relationships and multilateral bodies, including the Arctic Council, we will pursue arrangements that advance collective interests, promote shared Arctic state prosperity, protect the Arctic environment, and enhance regional security, and we will work toward U.S. accession to the United Nations Convention on the Law of the Sea (Law of the Sea Convention).

Our approach will be informed by the following guiding principles:

Safeguard Peace and Stability – Seek to maintain and preserve the Arctic region as an area free of conflict, acting in concert with allies, partners, and other interested parties. Support and preserve: international legal principles of freedom of navigation and overflight and other uses of the sea and airspace related to these freedoms, unimpeded lawful commerce, and the peaceful resolution of disputes for all nations.

Make Decisions Using the Best Available Information – Across all lines of effort, decisions need to be based on the most current science and traditional knowledge.[25]

Pursue Innovative Arrangements – Foster partnerships with the state of Alaska, Arctic states, other international partners, and the private sector to more efficiently develop, resource, and manage capabilities, where appropriate and feasible, to better advance our strategic priorities in this austere fiscal environment.

Consult and Coordinate with Alaska Natives – Engage in a consultation process with Alaska Natives, recognizing tribal governments' unique legal relationship with the United States and providing for meaningful and timely opportunity to inform Federal policy affecting Alaskan Native communities.[26]

For the main text of the document, see **Appendix D**.

January 2014 Implementation Plan for National Strategy for Arctic Region

On January 30, 2014, the Obama Administration released an implementation plan for the May 2013 national strategy for the Arctic region.[27] The plan states that it complements and builds upon existing initiatives by Federal, State, local, and tribal authorities, the private sector, and international partners, and focuses efforts where opportunities exist and action is most needed. The Implementation Plan reflects the reality of a changing Arctic environment and upholds national interests in safety, security, and environmental protection, and works with international partners to pursue global objectives of addressing climatic changes.

This Implementation Plan follows the structure and objectives of the Strategy's three lines of effort and is consistent with the guiding principles. The lines of effort of the Strategy and the Implementation Plan are as follows:

- Advance United States Security Interests
- Pursue Responsible Arctic Region Stewardship
- Strengthen International Cooperation

These lines of effort and guiding principles are meant to be implemented as a coherent whole.[28]

The plan also states:

Climate change is already affecting the entire global population, and Alaska residents are experiencing the impacts in the Arctic. To ensure a cohesive Federal approach, implementation activities must be aligned with the Executive Order on Preparing the United States for the Impacts of Climate Change[29] while executing the Strategy. In addition to the guiding principles, the following approaches are important in implementing the activities across all of the lines of effort:

- Foster Partnerships with Arctic Stakeholders. As outlined in the Strategy, all lines of effort must involve Arctic partners, particularly the State of Alaska and Alaska Natives in the Arctic region. Federal agencies, the State of Alaska, tribal communities, local governments, and academia will work with other nations, industry stakeholders, non-governmental organizations, and research partners to address emerging challenges and opportunities in the Arctic environment. The Federal Government should strive to maintain the free flow of communication and cooperation with the State of Alaska to support national priorities.
- Coordinate and Integrate Activities across the Federal Government. Multiple Federal bodies currently have authority for Arctic policy (e.g., the National Ocean Council (NOC), Arctic Policy Group, and Interagency Arctic Research Policy Committee (IARPC)). The National Security Council Staff will develop an Executive Order through the interagency process to maximize efficiency, align interagency initiatives, and create unity of effort among all Federal entities conducting activities in the Arctic.[30]

The plan outlines about 36 specific initiatives. For each, it presents a brief statement of the objective, a list of next steps to be taken, a brief statement about measuring progress in achieving the objective, and the names of the lead and supporting federal agencies to be involved.

The Arctic and the U.N. Convention on Law of the Sea (UNCLOS)[31]

Background to UNCLOS

In November 1994, the United Nations Convention on the Law of the Sea (UNCLOS) entered into force. This convention establishes a treaty regime to govern activities on, over, and under the world's oceans. It builds on the four 1958 law of the sea conventions and sets forth a framework for future activities in parts of the oceans that are beyond national jurisdiction.[32] The 1982 Convention and its 1994 Agreement relating to Implementation of Part XI of the Convention were transmitted to the Senate on October 6, 1994.[33] In the absence of Senate advice and consent to adherence, the United States is not a party to the convention and agreement.

Part VI of UNCLOS and Commission on Limits of Continental Shelf

Part VI of the convention, dealing with the Continental Shelf, and Annex II, which established a Commission on the Limits of the Continental Shelf, are most pertinent to the Arctic as it becomes more accessible ocean space, bordered by five coastal states.[34] The convention gives the coastal state sovereign jurisdiction over the resources, including oil and gas, of its continental shelf.[35] Under Article 76 of the convention, a coastal state with a broad continental margin may establish a shelf limit beyond 200 nautical miles. This jurisdiction is subject to the submission of the particulars of the intended limit and supporting scientific and technical data by the coastal state to the commission for review and recommendation.[36] The commission reviews the documentation and, by a two-thirds majority, approves its recommendations to the submitting state. Coastal states agree to establish the outer limits of their continental shelf, in accordance with this process and with their national laws. In instances of disagreement with the commission's recommendations, the coastal state may make a revised or new submission. The actions of the commission "shall not prejudice matters relating to delimitation of boundaries between States with opposite or adjacent coasts."[37] The "limits established by a coastal State on the basis of these recommendations shall be final and binding."[38]

U.S. Activities As a Non-Party to UNCLOS

As a non-party to the convention, the United States cannot participate as a member of the commission; it cannot submit a claim under Article 76. Over the years, however, it has submitted observations on submissions made by other states, requesting that those observations be made available online and to the commission. In addition, since 2001, the United States has gathered and analyzed data to determine the outer limits of its extended continental shelf. Starting in 2007, this effort became the Extended Continental Shelf Project, directed by an interagency task force under the lead of the Department of State.[39]

Some observers have suggested that a separate regime be negotiated to address the changing circumstances in the Arctic. They maintain that this phenomenon was not envisioned at the time the Law of the Sea Convention was negotiated. Still others suggest that the Arctic region above a certain parallel be designated a wilderness area, and they cite as precedent Article 4 of The Antarctic Treaty, under which any current claims to sovereign territory are frozen and

> No acts or activities taking place while the present Treaty is in force shall constitute a basis for asserting, supporting or denying a claim to territorial sovereignty in Antarctica or create any rights of sovereignty in Antarctica. No new claim, or enlargement of an existing claim, to territorial sovereignty in Antarctica shall be asserted while the present Treaty is in force.

Supporters of the Law of the Sea Convention maintain that changing circumstances in the Arctic strengthen their argument that the United States should become a party to the convention. In this way, they argue, the United States can be best situated to protect and serve its national interests, under both Article 76 and other parts of the convention.

The Administration's January 2014 implementation plan for its national strategy for the Arctic region (see discussion above) includes, as one of its 36 or so initiatives, one entitled "Accede to the Law of the Sea Convention." Under this initiative, the State Department and other federal agencies are to "Continue to seek the Senate's advice and consent to accede to the Law of the Sea Convention." The document states, "The Administration is committed, like the last three Administrations, to pursuing accession to the Convention on the Law of the Sea and will continue to place a priority on attaining Senate advice and consent to accession."[40] The Administration's January 2014 implementation plan for its national strategy for the Arctic region (see discussion above) includes, as one of its 36 or so initiatives, one entitled "Accede to the Law of the Sea Convention." Under this initiative, the State Department and other federal agencies are to "continue to seek the Senate's advice and consent to accede to the Law of the Sea Convention." The document states that "the Administration is committed, like the last three Administrations, to pursuing accession to the Convention on the Law of the Sea and will continue to place a priority on attaining Senate advice and consent to accession."[41]

ISSUES FOR CONGRESS

Climate Change and Loss of Arctic Sea Ice[42]

Record low extents of Arctic sea ice in 2012 and 2007 have focused scientific and policy attention on climate changes in the High North, and to the implications of projected ice-free[43] seasons in the Arctic within decades. The Arctic has been projected by several scientists to be ice-free in most late summers as soon as the 2030s.[44] This opens opportunities for transport through the Northwest Passage and the Northern Sea Route, extraction of potential oil and gas resources, and expanded fishing and tourism (**Figure 3**).

More broadly, physical changes in the Arctic include warming ocean, soil, and air temperatures; melting permafrost; shifting vegetation and animal abundances; and altered characteristics of Arctic cyclones. All these changes are expected to affect traditional livelihoods and cultures in the region and survival of polar bear and other animal populations, and raise risks of pollution, food supply, safety, cultural losses, and national security. Moreover, linkages ("teleconnections") between warming Arctic conditions and extreme events in the mid-latitude continents are increasingly evident, identified in such extreme events as the heat waves and fires in Russia in 2010; severe winters in the eastern United States and Europe in 2009/2010 and in Europe in 2011/2012;[45] and Indian summer monsoons and droughts. Hence, changing climate in the Arctic suggests important implications both locally and across the Hemisphere.

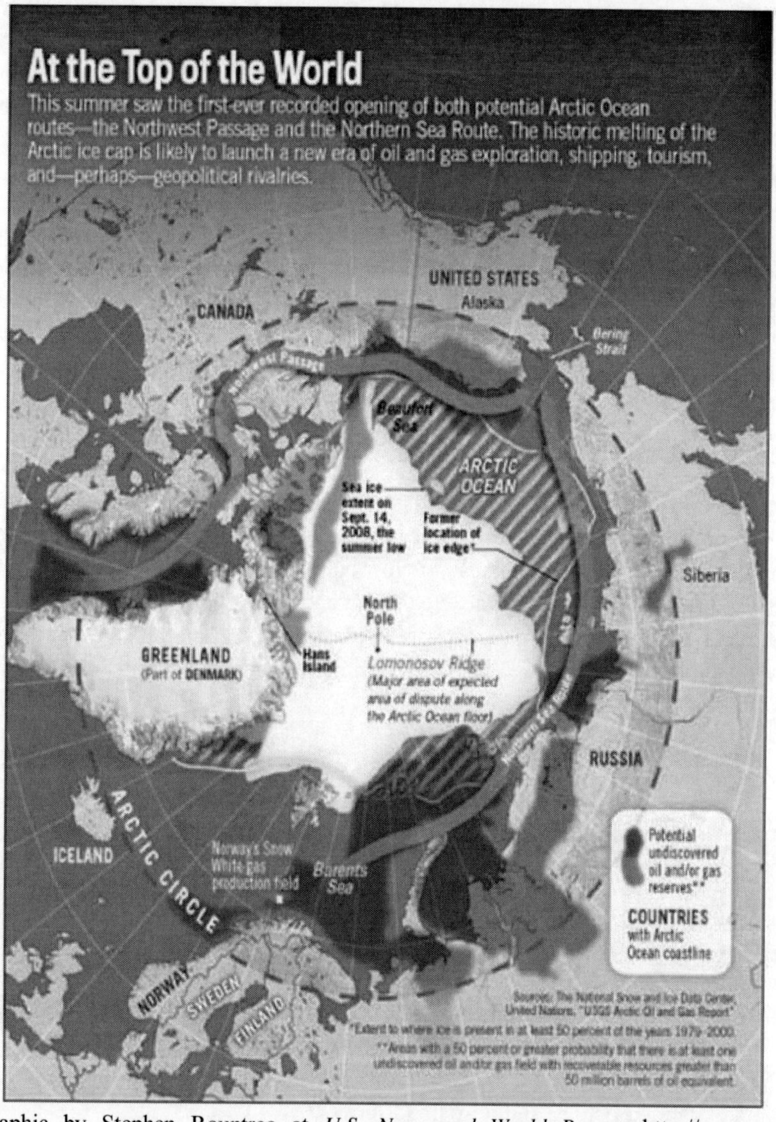

Source: Graphic by Stephen Rountree at *U.S. News and World Report*, http://www.usnews.com /articles/news/world/2008/10/09/global-warming-triggers-an-international-race-for-the-artic/photos/#1.

Figure 3. Arctic Sea Ice Extent in September 2008, Compared with Prospective Shipping Routes and Oil and Gas Resources.

Like the rest of the globe, temperatures in the Arctic have varied[46] but show a significant warming trend since the 1970s, and particularly since 1995.[47] The annual average temperature for the Arctic region (from 60° to 90° N) is now about 1.8° F warmer than the "climate normal" (the average from 1961 to 1990). Temperatures in October-November are now about

9° F above the seasonal normal. Scientists have concluded that most of the *global* warming of the last three decades is very likely caused by human-related emissions of greenhouse gases (GHG, mostly carbon dioxide); they expect the GHG-induced warming to continue for decades, even if, and after, GHG concentrations in the atmosphere have been stabilized. The extra heat in the Arctic is amplified by processes there (the "polar amplification") and may result in irreversible changes on human timescales.

The observed warmer temperatures along with rising cyclone size and strength in the Arctic have reduced sea ice extent, thickness, and ice that persists year-round ("perennial ice"); natural climate variability has likely contributed to the record low ice extents of 2007 and 2012. The 2007 minimum sea ice extent was influenced by warm Arctic temperatures and warm, moist winds blowing from the North Pacific into the central Arctic, contributing to melting and pushing ice toward and into the Atlantic past Greenland. Warm winds did not account for the near-record sea ice minimum in 2008.[48] In early August 2012, an unusually large storm with low pressure developed over the Arctic, helping to disperse the already weak ice into warmer waters and accelerating its melt rate. By August 24, 2012, sea ice extent had shrunk below the previous observed minimum of late September 2007.[49]

Modeling of GHG-induced climate change is particularly challenging for the Arctic, but it consistently projects warming through the 21st century, with annual average Arctic temperature increases ranging from +1° to +9.0° C (+2° to +19.0° F), depending on the GHG scenario and model used. While such warming is projected by most models throughout the Arctic, some models project slight cooling localized in the North Atlantic Ocean just south of Greenland and Iceland. Most warming would occur in autumn and winter, "with very little temperature change projected over the Arctic Ocean" in summer months.[50]

Due to observed and projected climate change, scientists have concluded that the Arctic will have changed from an ice-covered environment to a recurrent ice-free[51] ocean (in summers) as soon as the late 2030s. The character of ice cover is expected to change as well, with the ice being thinner, more fragile, and more regionally variable. The variability in recent years of both ice quantity and location could be expected to continue.

Territorial Claims and Sovereignty Issues[52]

Motivated in part by a desire to exercise sovereign control over the Arctic region's increasingly accessible oil and gas reserves (see "Oil, Gas, and Mineral Exploration"), the four Arctic coastal states other than the United States—Canada, Russia, Norway, and Denmark (of which Greenland is a territory)—are in the process of preparing territorial claims in the Arctic, including claims for expanded Exclusive Economic Zones (EEZs), for submission to the Commission on the Limits of the Continental Shelf. (As mentioned earlier—see "U.S. Activities As a Non-Party to UNCLOS"—the United States, as a non-party to UNCLOS, cannot participate as a member of the commission; it cannot submit a claim under Article 76. Over the years, however, it has submitted observations on submissions made by other states, requesting that those observations be made available online and to the commission. In addition, since 2001, the United States has gathered and analyzed data to determine the outer limits of its extended continental shelf.)

Russia has been attempting to chart the Arctic Ocean's enormous underwater Lomonosov Ridge in an attempt to show that it is an extension of Russia's continental margin. The

Russian claim to this ridge, if accepted, would reportedly grant Russia nearly one-half of the Arctic area; a 2001 claim submitted by Russia was rejected as insufficiently documented. Canada also claims a portion of the Lomonosov Ridge as part of its own underwater continental shelf.[53] In August 2007, a Russian submersible on a research expedition deposited an encased Russian Federation flag on the seabed of the presumed site of the North Pole. The action captured worldwide attention, but analysts note that it did not constitute an official claim to the territory and was therefore a purely symbolic act.

At a May 2008 meeting in Ilulissat, Greenland, the five Arctic coastal states reaffirmed their commitment to the UNCLOS legal framework for the establishment of extended continental shelf limits in the Arctic.[54] (For further discussion, see "Extent of the Continental Margin" in "Oil, Gas, and Mineral Exploration.")

In addition to this process, there are four unresolved Arctic territorial disputes:

- Scientists have forecast that in coming decades, global warming will reduce the ice pack in Canada's northern archipelago sufficiently to permit ships to use the trans-Arctic shipping route known as the Northwest Passage during the summer months (see "Commercial Sea Transportation"). The prospect of such traffic raises a major jurisdictional question. Ottawa maintains that such a passage would be an inland waterway, and would therefore be sovereign Canadian territory subject to Ottawa's surveillance, regulation, and control. The United States, the European Union, and others assert that the passage would constitute an international strait between two high seas.
- The United States and Canada are negotiating over a binational boundary in the Beaufort Sea.
- The United States and Russia in 1990 signed an agreement regarding a disputed area of the Bering Sea; the U.S. Senate ratified the pact the following year, but the Russian *Duma* has yet to approve the accord.
- Denmark and Canada disagree over which country has the territorial right to Hans Island, a tiny, barren piece of rock between Greenland and Canada's Ellesmere Island. Some analysts believe the two countries are vying for control over a future sea lane that might be created if the Arctic ice were to melt sufficiently to create a Northwest Passage. Others claim that the governments are staking out territorial claims in the event that future natural resource discoveries make the region economically valuable.[55]

In addition to these disputes, Norway and Russia had been at odds for decades over the boundary between the two in the so-called "Grey Zone" in the Barents Sea, an area believed to hold rich undersea deposits of petroleum. On September 15, 2010, Norwegian Prime Minister Jens Stoltenberg and Russian President Dmitry Medvedev signed an agreement in Murmansk, a Russian city near the Norwegian border. The accord awards roughly half of the 175,000-squarekilometer area to each country; it spells out fishing rights, and provides for the joint development of future oil and gas finds that straddle the boundary line. Some observers believe it is noteworthy that Russia would concede sovereignty over such a large, resource-rich territory to a small, neighboring country. But others have noted that Moscow may be hoping for Norwegian cooperation in developing offshore resources, and eventually in winning approval when Russia submits its Article 76 UNCLOS claim.[56]

In August 2010, Canadian Foreign Minister Lawrence Cannon announced a new "Statement of Canada's Arctic Policy," which reaffirmed the government's commitment to Canada's sovereignty in the region, to economic and social development, to environmental protection, and to empowerment of the peoples in the north. The statement also emphasized the government's intention to negotiate settlements to its disputes with the United States over the Beaufort Sea boundary, and with Denmark over Hans Island. Minister Cannon declared that "making progress on outstanding boundary issues will be a top priority."[57] Also, despite their dispute over Hans Island, Canada and Denmark have been working together on Arctic issues. In May 2010, the two countries' military chiefs of staffs signed a memorandum of understanding on Arctic Defense, Security, and Operational Cooperation, committing the two countries to "enhanced consultation, information exchange, visits, and exercises."[58]

Commercial Sea Transportation[59]

Background

The search for a shorter route from the Atlantic to Asia has been the quest of maritime powers since the Middle Ages. The melting of Arctic ice raises the possibility of saving several thousands of miles and several days of sailing between major trading blocs.[60] If the Arctic were to become a viable shipping route, the ramifications could extend far beyond the Arctic. For example, lower shipping costs could be advantageous for China (at least its northeast region), Japan, and South Korea because their manufactured products exported to Europe or North America could become less expensive relative to other emerging manufacturing centers in Southeast Asia, such as India.[61] Melting ice could potentially open up two trans-Arctic routes (see **Figure 3**):[62]

- **The Northern Sea Route** (NSR, a.k.a. the "Northeast Passage"), along Russia's northern border from Murmansk to Provideniya, is about 2,600 nautical miles in length. It was opened by the Soviet Union to domestic shipping in 1931 and to transit by foreign vessels in 1991. This route would be applicable for trade between northeast Asia (north of Singapore) and northern Europe. In the summer of 2013, about 50 laden cargo ships transited the NSR.[63]
- **The Northwest Passage** (NWP) runs through the Canadian Arctic Islands. The NWP actually consists of several potential routes. The southern route is through Peel Sound in Nunavut, which has been open in recent summers and contains mostly one-year ice. However, this route is circuitous, contains some narrow channels, and is shallow enough to impose draft restrictions on ships. The more northern route, through McClure Strait from Baffin Bay to the Beaufort Sea north of Alaska, is much more direct, and therefore more appealing to ocean carriers, but more prone to ice blockage.[64] The NWP is potentially applicable for trade between northeast Asia (north of Shanghai) and the northeast of North America, but may be less commercially viable than the NSR.[65] A ship carrying coal from western Canada to Finland became the first bulk carrier to transit the NWP in the summer of 2013.[66]

Destination Traffic, Not Trans-Arctic Traffic

Most cargo ship activity currently taking place in the Arctic is to transport natural resources from the Arctic or to deliver general cargo and supplies to communities and natural resource extraction facilities. Thus, cargo ship traffic in the Arctic presently is mostly regional, not trans-Arctic. While there has been a recent uptick in Arctic shipping activity, this activity has more to do with a spike in commodity prices than it does with the melting of Arctic ice. Even so, recent activity is less than it has been in the past. The NSR continues to account for the bulk of Arctic shipping activity.

Cruise Ship Activity

Considerable cruise ship activity takes place in Arctic waters. In the summer of 2007, three cruise ships reportedly sailed through the NWP from the Atlantic to Alaska's North Slope.[67] In August 2010, a cruise ship with over 150 passengers ran aground in the NWP. In the Barents Sea, there are regular cruise ships sailing to Svalbard. The inherent dangers for passenger ships in the Arctic have prompted calls for international regulations promoting the safety of cruise ships in the area. Some have suggested that cruise ships sail in pairs to provide assistance to one another, given the Arctic's remoteness and the difficulty land-based rescuers would have in reaching a vessel in distress.[68] Requiring that Arctic cruise vessels have ice-strengthened hulls and be equipped with enclosed lifeboats could be other safety requirements. In 2003, some Arctic cruise and tourist operators formed the Association of Arctic Expedition Cruise Operators (AECO) to establish agreed-upon safety and environmental protection guidelines, but this organization only covers the portion of the Arctic around Greenland, Svalbard, and Jan Mayen.[69]

Unpredictable Ice Conditions Hinder Trans-Arctic Shipping

Arctic waters do not necessarily have to be ice free to be open to shipping. Multiyear ice can be over 10 feet thick and problematic even for icebreakers, but one-year ice is typically 3 feet thick or less. This thinner ice can be more readily broken up by icebreakers or ice class ships (cargo ships with reinforced hulls and other features for navigating in ice-infested waters). However, more open water in the Arctic has resulted in another potential obstacle to shipping: unpredictable ice flows. In the NWP, melting ice and the opening of waters that were once covered with one-year ice has allowed blocks of multiyear ice from farther north, or icebergs from Greenland, to flow into potential sea lanes. The source of this multiyear ice is not predicted to dissipate in spite of climate change. Moreover, the flow patterns of these ice blocks are very difficult to predict, and they have floated into potential routes for shipping.[70] Thus, the lack of ice in potential sea lanes during the summer months can add even greater unpredictability to Arctic shipping. This is in addition to the extent of ice versus open water, which is also highly variable from one year to the next and seasonally.

The unpredictability of ice conditions is a major hindrance for trans-Arctic shipping in general, but can be more of a concern for some types of ships than it is for others. For instance, it would be less of a concern for cruise ships, which may have the objective of merely visiting the Arctic rather than passing through and could change their route and itinerary depending on ice conditions. On the other hand, unpredictability is of the utmost concern for container ships that carry thousands of containers from hundreds of different customers, all of whom expect to unload or load their cargo upon the ship's arrival at various ports as indicated on the ship's advertised schedule. The presence of even small blocks of ice or icebergs from a melting Greenland ice sheet requires slow sailing and could play havoc with schedules. Ships carrying a single commodity in bulk from one port to another for just one customer have more flexibility in terms of delivery windows, but would not likely risk an Arctic passage under prevailing conditions.

Ice is not the sole impediment to Arctic shipping. The region frequently experiences adverse weather, including not only severe storms, but also intense cold, which can impair deck machinery. During the summer months when sea lanes are open, heavy fog is common in the Arctic.

Commercial ships would face higher operating costs on Arctic routes than elsewhere. Ship size is an important factor in reducing freight costs. Many ships currently used in other waters would require two icebreakers to break a path wide enough for them to sail through; ship owners could reduce that cost by using smaller vessels in the Arctic, but this would raise the cost per container or per ton of freight.[71] Also, icebreakers or ice-class cargo vessels burn more fuel than ships designed for more temperate waters and would have to sail at slower speeds. The shipping season in the Arctic only lasts for a few weeks, so icebreakers and other special required equipment would sit idle the remainder of the year. None of these impediments by themselves may be enough to discourage Arctic passage but they do raise costs, perhaps enough to negate the savings of a shorter route. Thus, from the perspective of a shipper or a ship owner, shorter via the Arctic does not necessarily mean cheaper and faster.[72]

Basic Navigation Infrastructure Is Lacking

Considerable investment in navigation-related infrastructure would be required if trans-Arctic shipping were to become a reality. Channel marking buoys and other floating visual aids are not possible in Arctic waters because moving ice sheets will continuously shift their positions. Therefore, vessel captains would need to rely on marine surveys and ice charts. For some areas in the Arctic, however, these surveys and charts are out of date and not sufficiently accurate.[73] To remedy this problem, aviation reconnaissance of ice conditions and satellite images would need to become readily available for ship operators.[74] Ship-to-shore communication infrastructure would need to be installed where possible. Refueling stations may be needed, as well as, perhaps, transshipment ports where cargo could be transferred to and from ice-capable vessels at both ends of Arctic routes. Shipping lines would need to develop a larger pool of mariners with ice navigation experience. Marine insurers would need to calculate the proper level of risk premium for polar routes, which would require more detailed information about Arctic accidents and incidents in the past.

The U.S. Army Corps of Engineers, along with the state of Alaska, is studying the feasibility of a "deep-draft" port in the Arctic (accommodating ships with a draft of up to 35 feet). The northern and northwestern coastlines of Alaska are exceptionally shallow, generally limiting harbor and near shore traffic to shallow-draft barges. Coast Guard cutters and icebreakers have drafts of 35 to 40 feet while NOAA research vessels have drafts of 16 to 28 feet, so at present these vessels are based outside the Arctic and must sail considerable distances to reach Arctic duty stations. Supply vessels supporting offshore oil rigs typically have drafts over 20 feet. A deep-draft port could serve as a base of operations for larger vessels, facilitating commercial maritime traffic in the Arctic.[75] The study concluded that the existing harbors of Nome or Port Clarence on Alaska's west coast may be the most suitable for deepening because of their proximity to the Bering Strait and deeper water.[76]

The U.S. Committee on the Marine Transportation System, a Cabinet-level committee of federal agencies with responsibilities for marine transportation, identified and prioritized a list of infrastructure improvements for Arctic navigation in a 2013 report.[77] In the near term (two to three years), it prioritizes improvements to information infrastructure (weather forecasting, nautical charting, ship tracking) and emergency response capabilities for ships in distress.

Regulation of Arctic Shipping

Due to the international nature of the shipping industry, maritime trading nations have adopted international treaties that establish standards for ocean carriers in terms of safety, pollution prevention, and security. These standards are agreed upon by shipping nations through the International Maritime Organization (IMO), a United Nations agency that first met in 1959.[78]

Key conventions that the 168 IMO member nations have adopted include the Safety of Life at Sea Convention (SOLAS), which was originally adopted in response to the *Titanic* disaster in 1912 but has since been revised several times; the Prevention of Pollution from Ships (MARPOL), which was adopted in 1973 and modified in 1978; and the Standards for Training, Certification, and Watchkeeping for Seafarers (SCTW), which was adopted in 1978 and amended in 1995. It is up to ratifying nations to enforce these standards. The United States is a party to these conventions, and the U.S. Coast Guard enforces them when it boards and inspects ships and crews arriving at U.S. ports and the very few ships engaged in international trade that sail under the U.S. flag.

Like the United States, most of the other major maritime trading nations lack the ability to enforce these regulations as a "flag state" because much of the world's merchant fleet is registered under so-called "flags of convenience." While most ship owners and operators are headquartered in developed countries, they often register their ships in Panama, Liberia, the Bahamas, the Marshall Islands, Malta, and Cyprus, among other "open registries," because these nations offer more attractive tax and employment regulatory regimes. Because of this development, most maritime trading nations enforce shipping regulations under a "port state control" regime—that is, they require compliance with these regulations as a condition of calling at their ports. The fragmented nature of ship ownership and operation can be a further hurdle to regulatory enforcement. It is common for cargo ships to be owned by one company, operated by a second company (which markets the ship's space), and managed by a third (which may supply the crew and other services a ship requires to sail), each of which could be headquartered in different countries.

Arctic Guidelines

While SOLAS and other IMO conventions include provisions regarding the operation of ships in ice-infested waters, they are not specific to the polar regions. To supplement existing requirements, in December 2002, the IMO approved guidelines for ships operating in Arctic ice-covered waters.[79] These were only recommendations for ships operating in the Arctic, not requirements. They apply to passenger and cargo ships of 500 gross tons or more engaged in international voyages. They do not apply to fishing vessels, military vessels, pleasure yachts, and smaller cargo ships. The guidelines are intended to improve safety and prevent pollution in the Arctic, and they include provisions on ship construction, ship equipment related to navigation, and crew training and ship operation. The guidelines recommend that ships carry fully enclosed lifeboats or carry tarpaulins to cover their lifeboats. They recommend that each crew include at least one ice navigator with documented evidence of having completed an ice navigation training program. The IMO is in the process of drafting mandatory requirements for ships operating in the Arctic.[80]

Nations can enforce additional requirements on ships arriving at their ports or sailing through their coastal waters. For instance, U.S. Coast Guard regulations largely follow IMO conventions but mandate additional requirements in some areas. U.S. coastal states can require ships calling at their ports to take additional safety and pollution prevention safeguards. Canada and Russia have additional pollution regulations for Arctic waters exceeding MARPOL. The U.S. Coast Guard is seeking agreement with Russia to establish a vessel traffic separation scheme for the Bering Strait between Alaska and Russia, which now experiences over 300 transits per year.[81]

Oil, Gas, and Mineral Exploration[82]

Decreases in summer polar ice will likely allow for more options for exploring for oil and gas in certain offshore areas. Similarly, shrinking glaciers onshore could expose land containing economic deposits of gold, iron ore, or other minerals that were previously covered by glacial ice. Currently offshore areas encompassing more than 1 billion acres and more than 6,000 miles of coastline—more coastline than in the rest of the United States combined—are considered potential areas for offshore energy development. These Arctic regions include the Beaufort and Chukchi Seas, the Bering Sea, Cook Inlet, and the Gulf of Alaska.

Warming that causes permafrost to melt could pose challenges to onshore exploration activities because ground structures, such as pipelines and other infrastructure that depend on footings sunk into the permafrost for support, could become unstable. Despite the warming temperatures, however, exploration and development in the Arctic would still be subject to harsh conditions, especially in winter, which makes it costly and challenging to develop infrastructure necessary to produce, store, and transport oil, gas, and minerals from newly discovered deposits. In offshore areas, severe weather poses challenges to the ongoing operation of federally regulated operations involving six producing wells.[83]

Shrinking sea ice cover in the Arctic has also intensified interest in surveying and mapping the continental margins of countries with lands in the Arctic. Delineating the extent of the continental margins beyond the 200 nautical mile Exclusive Economic Zone (EEZ) could lead to considering development on substantial amounts of submerged lands.[84] Mapping projects are underway, by individual countries and through cooperative government studies, to support national claims to submerged lands which may contain large amounts of oil, natural gas, methane hydrates, or minerals.

One historical determinant of energy resource management in the Arctic has been federal moratoriums enacted by Congress on an annual basis (as part of DOI appropriations acts) and issued by the President through treaties or executive orders.[85] While not affecting all Alaska operations, these moratoriums effectively banned federally regulated planning and permitting in the Bristol Bay area of the North Aleutian Basin. Since Congress allowed most statutory bans in the region to expire in 2004,[86] federal policy on development rights and permits governing access to the Arctic for drilling has been in flux. In 2010, President Obama reinstated a moratorium in the North Aleutian Basin, withdrawing acreage located in Bristol

Bay from eligibility for oil and gas leasing until after 2017.[87] Pursuant to this executive ban, gathering geological and geophysical data and other development-related activities are prohibited in Bristol Bay.

Federal officials report regional operations related to the six producing wells are stable[88] and that plans are unfolding for oil and gas leasing activity in the future.[89] Prompted by claims voiced by some Alaska state officials and energy, trade, and shipping groups favoring greater geographical diversity for federally regulated offshore energy projects, President Obama and many in Congress have expressed interest in expanding America's ocean energy portfolio in the region.[90] As part of the Administration's strategy to enhance domestic energy production, DOI approved a program referred to as the Five Year Program announcing regulatory actions in offshore areas from 2012 through 2017.[91] This program was formulated in part by lessons learned in the aftermath of the *Deepwater Horizon* events in the Gulf of Mexico.[92] While the program defers activities in the Bristol Bay (pursuant to the executive moratorium), it proposes three lease sales elsewhere in the region: in the Chukchi Sea (2016), Cook Inlet (2016), and the Beaufort Sea (2017).

One example of current offshore permitting activity in the region includes a subsidiary of Shell Oil Company[93] gaining permits for exploratory oil and natural gas drilling operations in the Chukchi and Beaufort Seas during the 2012 drilling season.[94] Public attention to these Shell permits peaked during reports of a December 31, 2012, incident involving a drillship named the *Kulluk*, owned by Shell. Media reports that the drillship ran aground on the shoreline of Sitkalidak Island, AK, sparked attention to safety issues.[95] As part of a response to the grounding, federal safety officials commenced an investigation of the incident.[96] Industry commentators, including Randall Luthi, President of the National Ocean Industries Association (NOIA), noted that drilling in U.S. waters can be achieved safely and that this incident should not be a stumbling block to further development in the region.[97] Other commentators countered that developing Arctic resources is risky and investigating this incident offers guidance for future permitting in the region.[98]

The future of offshore drilling in the region is somewhat dependent on the outcome of judicial proceedings that govern aspects of Shell's permitting process.[99] The future timetable for Shell's permits might also depend on DOI regulatory actions focusing on, among other issues, the *Arctic Challenger* (an Arctic Containment System, ACS), a key component of Shell's drilling plan.[100] As part of listing various operational and management issues in the investigatory report described above, DOI announced further studies are warranted prior to approving future Shell drilling operations. Specifically, in terms of fabricating and testing certain safety systems and establishing the scope of its operational plans, DOI investigators found Shell lacked adequate preparation and Shell's timelines did not allow for full certification of its ACS.[101] Based on the information available to date from the courts and from DOI on the status of the Shell permitting process, it remains to be seen how permitting for drilling in the Arctic might proceed, if at all, with respect to the 2013 drilling season.

Oil and Gas

A primary driver for the increased interest in exploring for oil and gas offshore in the Arctic is the shrinking Arctic ice cap, or conversely, the growing amount of ice-free ocean in the summertime. Reduced sea ice in the summer means that ships towing seismic arrays[102] can explore previously inaccessible regions of the Arctic Ocean, Chukchi Sea, Beaufort Sea, and other offshore regions for longer periods of time without as much risk of colliding with floating sea ice. Less sea ice over longer periods compared to previous decades also means that the seasonal window for offshore drilling in the Arctic remains open longer in the summer and increases the chances for making a discovery.

In addition to the improved access to larger portions of the Arctic afforded by shrinking sea ice, interest in Arctic oil and gas was fueled by a 2008 U.S. Geological Survey (USGS) appraisal of undiscovered oil and gas north of the Arctic Circle.[103] The USGS asserted that the "extensive Arctic continental shelves may constitute the geographically largest unexplored prospective area for petroleum remaining on Earth."[104] In the report, the USGS estimates that 90 billion barrels of oil, nearly 1,700 trillion cubic feet of natural gas, and 44 billion barrels of natural gas liquids may remain to be discovered in the Arctic. An article published in Science magazine indicated that 30% of the world's undiscovered natural gas and 13% of the world's undiscovered oil may be found north of the Arctic Circle.[105]

Despite the warming trend in the Arctic, a discovery of new oil and gas deposits far from existing storage, pipelines, and shipping facilities cannot be developed until infrastructure is built to extract and transport the petroleum. Other areas with oil and gas potential, such as offshore east Greenland, may take even longer to develop because they are still covered by substantial summer sea ice that would hinder exploration and development unless more ice melts.[106] Further, in 2010 Secretary of the Interior Salazar asked the USGS to assess the science needs that would inform the Obama Administration's consideration of how to develop oil and gas resources in the Arctic.[107] The USGS released a report in response that identified some overarching themes, which include the need for (1) large-scale synthesis of data and information, (2) enhanced dialog and collaborative science planning, and (3) a more transparent and inclusive planning and decision-making process.[108]

Extent of the Continental Margin

Increased interest in developing offshore resources in the Arctic has sparked efforts by nations bordering the Arctic Ocean to map the extent of their continental margins beyond the 200-mile EEZ limit. Under UNCLOS, nations can submit a claim to submerged lands if they demonstrate that their continental margin extends beyond the 200-mile limit according to the definition in Article 76. Under Article 76, the extent of the continental margin beyond the 200-mile limit depends on the position of the foot of the continental slope, the thickness of sediments, and the depth of water. Also, the continental margin could include geologic features that extend from the continent out to sea, which may include undersea ridges continuing for hundreds of miles offshore.

As mentioned earlier—see "U.S. Activities As a Non-Party to UNCLOS"—the United States, as a non-party to UNCLOS, cannot participate as a member of the Commission on the Limits of the Continental Shelf; it cannot submit a claim under Article 76. Over the years, however, it has submitted observations on submissions made by other states, requesting that those observations be made available online and to the commission. In addition, since 2001 the United States has gathered and analyzed data to determine the outer limits of its extended continental shelf.

Arctic border countries have begun the complex investigations needed to support claims for an extended continental shelf in the Arctic. Claims have already been submitted by several countries, including the Russian Federation, which submitted its UNCLOS claim to a portion of the Arctic continental shelf in 2001.[109] Russia made claims to the Lomonosov Ridge, an undersea feature spanning the Arctic from Russia to Canada, as an extension of its continental margin. The submission demonstrated Russia's bid to extend activities in Arctic regions. The United States has started to gather and analyze data through an initiative called the Extended Continental Shelf Project.[110] In this effort, the United States is working closely with Canada to prepare and present Canada's submission to the Commission on the Limits of the Continental Shelf.

Canada and the United States share overlapping regions of the seabed as part of the extended continental margin of both nations. Both countries have conducted research singly and jointly to map the extended continental shelf.[111] Consistent with past U.S. directives addressing the extended continental shelf and boundary issues in the Arctic,[112] the Obama Administration has undertaken an initiative to improve the delivery of relevant scientific information to officials responsible for energy development decisions in the Arctic.[113] This initiative is part of a broader response to USGS recommendations that more dialogue and collaborative science planning occur between and among the scientific community and federal agencies involved in Arctic oil and gas development decisions.[114] These USGS recommendations stem from earlier USGS findings that most of the potential oil and gas resources estimated for the Arctic are likely to exist within already agreed-upon territorial boundaries.[115] (For further discussion, see "Territorial Claims and Sovereignty Issues.")

Minerals

A warming Arctic means new opportunities and challenges for mineral exploration and development. Receding glaciers expose previously ice-covered land that could host economic mineral deposits that were previously undetectable and un-mineable below the ice. Longer summers would also extend exploration seasons even for areas not currently ice-covered but which are only accessible for ground surveys during the warmer months. In some parts of the Arctic, such as Baffin Island, Canada, less sea ice allows ships to transport heavy equipment to remote locations, and to convey ore from mines to the market further south. Some railway and mining operators are considering developing railroads and other infrastructure to transport ore year-round.[116] As with onshore oil and gas development, however, mining infrastructure that depends on footings sunk into permafrost could become unstable if the permafrost melts in response to warmer temperatures. Also, as with oil and gas development, mineral deposits that may be technically recoverable with current technology may not be economically profitable.

Some industry commentators suggest that mining might offer better long-term economic development opportunities compared to oil and gas development because of a larger permanent workforce and project lifelines of several decades.[117] Similar to oil and gas, however, industry observers note that uncertainties and knowledge gaps exist in the understanding of environmental change in the Arctic, and how to deal with the risks of significant Arctic industrial activity.[118]

One important part of the current infrastructure in the Arctic that supports oil, gas, and mineral development is the construction and use of ice roads—built and used during the winter—but which are not passable during the warmer months. Warmer temperatures are shortening the ice road transport seasons and creating transportation challenges. For example, the opening date for tundra roads in northern Alaska usually occurred in early November prior to 1991 and has shifted to January in recent years.[119]

Oil Pollution and Pollution Response[120]

Oil Pollution Implications of Arctic Change

Climate change impacts in the Arctic, particularly the decline of sea ice and retreating glaciers, have stimulated human activities in the region, many of which have the potential to create oil pollution. A primary concern is the threat of a large oil spill in the area. Although a major oil spill has not occurred in the Arctic region,[121] recent economic activity, such as oil and gas exploration and tourism (cruise ships), increases the risk of oil pollution (and other kinds of pollution) in the Arctic. Significant spills in high northern latitudes (e.g., the 1989 *Exxon Valdez* spill in Alaska and spills in the North Sea) suggest that the "potential impacts of an Arctic spill are likely to be severe for Arctic species and ecosystems."[122]

Risk of Oil Pollution in the Arctic

A primary factor determining the risk of oil pollution in the Arctic is the level and type of human activity being conducted in the region. Although climate changes in the Arctic are expected to increase access to natural resources and shipping lanes, the region will continue to present logistical challenges that may hinder human activity in the region. For example (as discussed in another section of this report),[123] the unpredictable ice conditions may discourage trans-Arctic shipping. If trans-Arctic shipping were to occur on a frequent basis, it would represent a considerable portion of the overall risk of oil pollution in the region. In recent decades, many of the world's largest oil spills have been from oil tankers, which can carry millions of gallons of oil.[124]

Although the level of trans-Arctic shipping is uncertain, many expect oil exploration and extraction activities to intensify in the region.[125] Oil well blowouts from offshore oil extraction operations have been a source of major oil spills, eclipsing the largest tanker spills. The largest unintentional oil spill in recent history was from the 2010 *Deepwater Horizon* incident in the Gulf of Mexico.[126] During that incident, the uncontrolled well released (over an 84-day period) approximately 200 million gallons of crude oil into the Gulf.[127] The second-largest unintentional oil spill in recent history—the *IXTOC I*, estimated at 140 million gallons—was due to an oil well blowout in Mexican Gulf Coast waters in 1979.[128]

Until the 2010 *Deepwater Horizon* incident, the spill record for offshore platforms in U.S. federal waters had shown improvement from prior years.[129] A 2003 National Research Council (NRC) study of oil and gas activities on Alaska's North Slope stated "the conclusion of these analyses is that blowouts that result in large spills are unlikely."[130] Similar conclusions were made in federal agency documents regarding deepwater drilling in the Gulf of Mexico before the 2010 *Deepwater Horizon* event.[131] Some would likely contend that the underlying analyses behind these conclusions should be adjusted to account for the 2010 Gulf oil spill. However, others may argue that the proposed activities in U.S. Arctic waters present less risk of an oil well blowout than was encountered by the *Deepwater Horizon* drill rig, because the proposed U.S. Arctic operations would be in shallower waters (150 feet) than the deepwater well (approximately 5,000 feet) that was involved in the 2010 Gulf oil spill. In addition, Shell Oil has stated that the pressures in the Chukchi Sea (the location of Shell's recent interest) would be two to three times less than they were in well involved in the 2010 Gulf oil spill.[132] Regardless of these differences, even under the most stringent control systems, some oil spills and other accidents are likely to occur from equipment failure or human error.

Potential Impacts

No oil spill is entirely benign. Even a relatively minor spill, depending on the timing and location, can cause significant harm to individual organisms and entire populations. Regarding aquatic spills, marine mammals, birds, bottom-dwelling and intertidal species, and organisms in early developmental stages—eggs or larvae—are especially vulnerable. However, the effects of oil spills can vary greatly. Oil spills can cause impacts over a range of time scales, from only a few days to several years, or even decades in some cases.

Conditions in the Arctic may have implications for toxicological effects that are not yet understood. For example, oil spills on permafrost may persist in an ecosystem for relatively long periods of time, potentially harming plant life through their root systems. Moreover, little is known about the effects of oil spills on species that are unique to the Arctic, particularly, species' abilities to thrive in a cold environment and the effect temperature has on toxicity.[133]

The effects of oil spills in high latitude, cold ocean environments may last longer and cause greater damage than expected. Some recent studies have found that oil spills in lower latitudes have persisted for longer than initially expected, thus raising the concern that the persistence of oil in the Arctic may be understated. In terms of wildlife, population recovery may take longer in the Arctic because many of the species have longer life spans and reproduce at a slower rate.[134]

Response and Cleanup Challenges in the Arctic Region

Climate changes in the Arctic are expected to increase human activities in the region, many of which impose a risk of oil pollution, particularly from oil spills. Conditions in the Arctic region impose unique challenges for personnel charged with (1) oil spill response, the process of getting people and equipment to the incident, and (2) cleanup duties, either recovering the spilled oil or mitigating the contamination so that it poses less harm to the ecosystem. These challenges may play a role in the policy development for economic activities in the Arctic.

Spill Response Challenges

Response time is a critical factor for oil spill recovery. With each hour, spilled oil becomes more difficult to track, contain, and recover, particularly in icy conditions, where oil can migrate under or mix with surrounding ice.[135] Most response techniques call for quick action, which may pose logistical challenges in areas without prior staging equipment or trained response professionals. Many stakeholders are concerned about a "response gap" for oil spills in the Arctic region.[136] A response gap is a period of time in which oil spill response activities would be unsafe or infeasible. The response gap for the northern Arctic latitudes is likely to be extremely high compared to other regions.[137]

According to the former Commander of the 17th Coast Guard District (Alaska), "we are not prepared for a major oil spill [over 100,000 gallons] in the Arctic environment. The Coast Guard has no offshore response capability in Northern or Western Alaska."[138] The transportation infrastructure along Alaska's northern coast poses challenges for oil spill responders. The Coast Guard has no designated air stations north of Kodiak, AK, which is almost 1,000 miles from the northernmost point of land along the Alaskan coast in Point Barrow, AK.[139] Although some of the communities have airstrips capable of landing cargo planes, no roads connect these communities.[140] Vessel infrastructure is also limited. The nearest major port is in the Aleutian Islands, approximately 1,300 miles from Point Barrow. A 2010 Government Accountability Office (GAO) report identified further logistical obstacles that would hinder an oil spill response in the region, including "inadequate" ocean and weather information for the Arctic and technological problems with communications.[141]

The history of oil spills and response in the Aleutian Islands raises concerns for potential spills in the Arctic region:

> The past 20 years of data on response to spills in the Aleutians has also shown that almost no oil has been recovered during events where attempts have been made by the responsible parties or government agencies, and that in many cases, weather and other conditions have prevented any response at all.[142]

Oil Spill Cleanup Challenges

The behavior of oil spills in cold and icy waters is not well understood.[143] Cleaning up oil spills in ice-covered waters will be more difficult than in other areas, primarily because effective strategies have yet to be developed. Natural oil seeps, which are a major source of oil in the arctic environment, may offer opportunities for studying the behavior of oil.[144]

The Arctic conditions present several hurdles to oil cleanup efforts. In colder water temperatures, there are fewer organisms to break down the oil through microbial degradation. Oil evaporates at a slower rate in colder water temperatures. Although slower evaporation may allow for more oil to be recovered, evaporation removes the lighter, more toxic hydrocarbons that are present in crude oil.[145] The longer the oil remains in an ecosystem, the more opportunity there is for exposure. Oil spills may get trapped in ice, evaporating only when the ice thaws. In some cases, oil could remain in the ice for years.

Icy conditions enhance emulsification—the process of forming different states of water in oil, often described as "mousse." Emulsification creates oil cleanup challenges by increasing (1) the volume of the oil/water mixture and (2) the mixture's viscosity (resistance to flow). The latter change creates particular problems for conventional removal and pumping cleanup methods.[146] Moreover, two of the major non-mechanical recovery methods—in-situ burning and dispersant application—may be limited (or "precluded") by the Arctic conditions and lack of logistical support: aircraft, vessels, and other infrastructure.[147]

Existing Policy Framework

Considering both the recent increase in human activity in the region (and expectation of further interest) and the response and recovery challenges that an oil spill would impose in Arctic waters, many would assert that the region warrants particular attention in terms of governance. However, the existing framework for international governance of maritime operations in the Arctic region lacks legally binding requirements. While the Safety of Life at Sea Convention (SOLAS) and other International Maritime Organization (IMO) conventions include provisions regarding ships in icy waters, the provisions are not specific to the polar regions. Although the IMO has "Guidelines for Ships Operating in Arctic," a 2009 NOAA report described the non-binding IMO provisions as "inconsistent with the hazards of Arctic navigation and the potential for environmental damage from such an incident."[148]

Fisheries[149]

Large commercial fisheries exist in the Arctic, including in the Barents and Norwegian Seas north of Europe, the Central North Atlantic off of Greenland and Iceland, and the Newfoundland and Labrador Seas off of northeastern Canada.[150]

In the 110th Congress, P.L. 110-243 stated that "the United States should initiate international discussions and take necessary steps with other Arctic nations to negotiate an agreement or agreements for managing migratory, transboundary, and straddling fish stocks in the Arctic Ocean and establishing a new international fisheries management organization or organizations for the region." In response to this, the United States participated in meetings with other Arctic nations at the March 2009 U.N. Food and Agriculture Organization Committee on Fisheries meetings in Rome. These international discussions are continuing. No additional legislation has been enacted related to U.S. fisheries in the Arctic.

On February 5, 2009, the North Pacific Fishery Management Council, acting under the authority of the Magnuson-Stevens Fishery Conservation and Management Act, unanimously approved the Council's draft Arctic Fisheries Management Plan and approved the preferred alternative, which would (1) close the Arctic to commercial fishing until information improves so that fishing can be conducted sustainably and with due concern to other ecosystem components; (2) determine the fishery management authorities in the Arctic and provide the Council with a vehicle for addressing future management issues; and (3) implement an ecosystem-based management policy that recognizes the resources of the U.S. Arctic and the potential for fishery development that might affect those resources, particularly in the face of a changing climate.

On November 3, 2009, the National Marine Fisheries Service (Department of Commerce) issued a final rule that implemented the North Pacific Council's Fishery Management Plan for Fish Resources of the Arctic Management Area, as well as Amendment 29 to the Fishery Management Plan for Bering Sea/Aleutian Islands King and Tanner Crabs. This action followed the North Pacific Council's recommendations establishing a basis for sustainable management of commercial fishing in the Arctic Management Area and moving the northern boundary of the crab fishery south of the Arctic Management Area to Bering Strait.[151] In October 2009, an international symposium focused on conservation of fish stocks and the potential for future fisheries management in the Arctic Ocean.[152] The National Marine Fisheries Service maintains an "Arctic Fisheries" web page summarizing activates related to fisheries.[153]

Protected Species[154]

Concern over development of the Arctic relates to how such development might affect threatened and endangered species. Under the Endangered Species Act (ESA, 16 U.S.C. §§1531-1543), the polar bear was listed as threatened on May 15, 2008. The failure by the Fish and Wildlife Service (FWS) to make a 90-day finding on a 2008 petition to list Pacific walrus led to submission of 60- days' notice of a future citizen suit. However, eventually walruses were listed as candidate species under ESA;[155] this status means that federal agencies carrying out actions that may affect the species must confer with FWS though they are not necessarily obliged to modify their actions. Both polar bears and walruses are heavily dependent during their life cycles on thick sea ice, making them especially susceptible to the shrinking Arctic ice cap.

On December 30, 2008, the National Marine Fisheries Service (NMFS) determined that a listing of ribbon seal as threatened or endangered was not warranted.[156] On October 22, 2010, NMFS listed the southern distinct population segment (DPS) of spotted seals as threatened.[157] Listing of two other DPS (Okhotsk and Bering Sea) had earlier been determined to not be warranted.[158] On December 10, 2010, NMFS proposed that (1) four subspecies of ringed seal be listed as threatened,[159] and (2) that two DPS of one subspecies of bearded seal be listed as threatened.[160]

In either terrestrial or marine environments, the extreme pace of change makes a biological response many times more difficult. For species with adaptations for a specific optimum temperature for egg development, or production of young timed to match the availability of a favored prey species, or seed dispersal in predictable fire regimes, etc., evolutionary responses may well not keep pace with the rate of change.[161] While species of plants and animals farther south might migrate, drift, or be transplanted from warming habitats to more northerly sites that may continue to be suitable,[162] once a terrestrial species reaches the Arctic Ocean, it is very literally at the end of the line. No more northern or colder habitat is available.

The Marine Mammal Protection Act (MMPA; 16 U.S.C. §§1361 et seq.) protects whales, seals, walruses, and polar bears. The MMPA established a moratorium on the "taking" of marine mammals in U.S. waters and by U.S. nationals on the high seas, including the Arctic. The MMPA protects marine mammals from "clubbing, mutilation, poisoning, capture in nets, and other human actions that lead to extinction." Under the MMPA, the Secretary of Commerce, acting through National Marine Fisheries Service (NMFS), is responsible for the conservation and management of whales and seals. The Secretary of the Interior, acting through the Fish and Wildlife Service (FWS), is responsible for walruses and polar bears.[163] Despite the MMPA's general moratorium on taking, the MMPA allows U.S. citizens to apply for and obtain authorization for taking small numbers of mammals incidental to activities other than commercial fishing (e.g., offshore oil and gas exploration and development) if the taking would have only a negligible impact on any marine mammal species or stock, provided that monitoring requirements and other conditions are met.

Indigenous People Living in the Arctic[164]

People have been living in the Arctic for thousands of years, and indigenous peoples developed highly specialized cultures and economies based on the physical and biological conditions of the long-isolated region. However, with trade, the influx of additional populations especially since the 19th century, and ongoing physical changes in the Arctic, indigenous populations have already experienced substantial change in their lifestyles and economies. Over the past two decades, greater political organization across indigenous populations has increased their demands for international recognition and broader rights, as well as attention to the economic, health, and safety implications of climate change in the North.

Background

Seven of the eight Arctic nations have indigenous peoples,[165] whose predecessors were present in parts of the Arctic over 10,000 years ago, well before the arrival of peoples with European backgrounds.[166] Current Arctic indigenous peoples comprise dozens of diverse cultures and speak dozens of languages from eight or more non-Indo-European language families.[167]

Before the arrival of Europeans, Arctic indigenous peoples lived in economies that were chiefly dependent, in varying proportions, on hunting land and marine mammals, catching salt- and fresh-water fish, herding reindeer (in Eurasia), and gathering, for their food, clothing, and other products.[168] Indigenous peoples' interaction with and knowledge of Arctic wildlife and environments has developed over millennia and is the foundation of their cultures.[169]

The length of time that Arctic indigenous peoples were in contact with Europeans varied across the Arctic. As recorded by Europeans, contact began as early as the 9th century CE, if not before, in Fennoscandia[170] and northwestern Russia, chiefly for reasons of commerce (especially furs); it progressed mostly west-to-east across northern Asia, reaching northeastern Arctic Asia by the 17th century.[171] North American Arctic indigenous peoples' contact with Europeans started in Labrador in the 16th century and in Alaska in the 18th century, and was not completed until the early 20th century.[172] Greenland's indigenous peoples first saw European-origin peoples in the late 10th century, but those Europeans died out during the 15th or 16th century and Europeans did not return permanently until the 18th century.[173]

Contact led to significant changes in Arctic indigenous economies, political structures, foods, cultures, and populations, starting especially in the 20th century. For example, life expectancy among Alaska Natives has increased from 47 years in 1950 to over 69 years in 2000 (though it still lags behind that of U.S. residents overall, at 77 years).[174]

Also, at present, most Arctic indigenous peoples have become minorities in their countries' Arctic areas, except in Greenland and Canada. (One source estimates that, around 2003, about 10% of an estimated 3.7 million people in the Arctic were indigenous.)[175] While many Arctic indigenous communities remain heavily dependent on hunting, fishing, and herding and are more likely to depend on traditional foods than non-indigenous Arctic inhabitants,[176] there is much variation. Most Arctic indigenous people may no longer consume traditional foods as their chief sources of energy and nutrition.[177] Major economic change is also relatively recent but ongoing.[178] Many Arctic indigenous communities have developed a mixture of traditional economic activities and wage employment.[179] The economics of subsistence and globalization will be key factors in the effects of climate change on Arctic indigenous peoples, and on their reactions to Arctic climate change.

Arctic indigenous peoples' current political structures vary, as do their relationships with their national governments. Some indigenous groups govern their own unique land areas within the national structure, as in the United States and Canada; others have special representative bodies, such as the Saami parliaments in Norway, Finland, and Sweden;[180] a few areas have general governments with indigenous majorities, such as Greenland (a member country of Denmark), Nunavut territory in Canada, and the North Slope and Northwest Arctic boroughs in Alaska.[181] Control of land, through claims and ownership, also varies among Arctic indigenous peoples, as do rights to fishing, hunting, and resources.[182] Arctic indigenous peoples' political relationships to their national and local governments, and their ownership or claims regarding land, are also significant factors in the responses to Arctic climate change by the indigenous peoples and by Arctic nations' governments.

Effects of Climate Change

Arctic climate change is expected to affect the economies, population, subsistence, health, infrastructure, societies, and cultures of Arctic indigenous peoples. Changes in sea ice and sea level, permafrost, tundra, weather, and vegetation distributions, as well as increased commercial shipping, mineral extraction, and tourism, will affect the distribution of land and sea mammals, of freshwater and marine fish, and of forage for reindeer. These will in turn affect traditional subsistence activities and related indigenous lifestyles.[183] Arctic indigenous peoples' harvesting of animals is likely to become riskier and less predictable, which may increase food insecurity, change diets, and increase dependency on outside, non-traditional foods.[184] Food cellars in many locations have thawed during summers, threatening food safety. Related health risks of diabetes, obesity, and mental illness have been associated with these changes.[185]

Sea, shoreline ice, and permafrost changes have damaged infrastructure and increased coastal and inland erosion, especially in Alaska, where GAO found in 2003 that "coastal villages are becoming more susceptible to flooding and erosion caused in part by rising temperatures."[186] In response, Congress funded the U.S. Army Corps of Engineers to conduct a Baseline Erosion Assessment that identified and prioritized among the 178 communities identified at risk from erosion.[187] (Risks from flooding were not examined.) GAO concluded in 2009 that many Native villages must relocate, but even those facing imminent threats have been impeded by various barriers, including difficulties identifying appropriate new sites, piecemeal programs for state and federal assistance, and obstacles to eligibility for certain federal programs.[188] The Alaska Federation of Natives placed among its 2010 federal priorities a request to Congress to mitigate flooding and erosion in Alaska Native villages and to fund relocation of villages where necessary.[189] However, "the cost is extraordinary," acknowledges Senator Lisa Murkowski.[190]

Oil, gas, and mineral exploration and development are expected to increase, as are other economic activities, such as forestry and tourism, and these are expected to increase economic opportunities for all Arctic residents, including indigenous peoples.[191] Pressures to increase participation in the wage economy, however, may speed up changes in indigenous cultures. Increased economic opportunities may also lead to a rise in the non-indigenous population, which may further change the circumstances of indigenous cultures. Some representatives of Arctic indigenous people have related a "conflicting desire between combating climate change and embracing the potential for economic growth through foreign investment."[192]

Although important advances in public health have occurred in indigenous communities over past decades, some health problems may increase with continued Arctic climate change. Economic development may exacerbate Arctic pollution problems, including higher exposure to mercury, air pollution, and food contamination. The influx and redistribution of contaminants in the air, oceans, and land may change in ways that are now poorly understood.[193] Warmer temperatures and longer warm seasons may increase insect- and wildlife-borne diseases.[194] Climate change may lead to damage to water and sanitation systems, reducing protection against waterborne diseases.[195] Changes in Arctic indigenous cultures may increase mental stress and behavioral problems.[196]

The response to climate change by Arctic indigenous peoples has included international activities by Arctic indigenous organizations and advocacy before their national governments. As one report noted, "the rise of solidarity among indigenous peoples organizations in the region is surely a development to be reckoned with by all those interested in policy issues in the Arctic."[197] Six national or international indigenous organizations are permanent participants of the Arctic Council, the regional intergovernmental forum.[198] Due in part to advocacy by Arctic indigenous people, the United Nations General Assembly adopted in 2007 the Declaration on the Rights of Indigenous Peoples.[199] In April 2009, the Inuit Circumpolar Council (an organization of Inuit in the Arctic regions of Alaska, Canada, Greenland, and Russia) hosted in Alaska the worldwide "Indigenous Peoples Global Summit on Climate Change."[200] The conference report, forwarded to the Copenhagen Conference of the Parties of the U.N. Framework Convention on Climate Change (December 2009), noted "accelerating" climate change caused by "unsustainable development" and, among several recommendations, called for a greater indigenous role in national and international decisions on climate change, including a greater role for indigenous knowledge in climate change research, monitoring, and mitigation.[201]

Polar Icebreaking[202]

The Coast Guard's polar icebreakers perform a variety of missions supporting U.S. interests in the Arctic (and Antarctic), including the following:

- conducting and supporting U.S. scientific research in the Arctic;[203]
- defending U.S. sovereignty in the Arctic by helping to maintain a presence in the region;
- defending other U.S. interests in the Arctic, including economic interests relating to the U.S. exclusive economic zone (EEZ) north of Alaska;
- monitoring sea traffic in the Arctic, including ships bound for the United States; and
- conducting other typical Coast Guard missions (such as search and rescue, law enforcement, and protection of marine resources) in Arctic waters, including U.S. territorial waters north of Alaska.

The Coast Guard's strategy document for the Arctic region, released on May 21, 2013, states that "The United States must have adequate icebreaking capability to support research that advances fundamental understanding of the region and its evolution," and that "The Nation must also make a strategic investment in icebreaking capability to enable access to the high latitudes over the long-term."[204]

The Coast Guard's two existing heavy polar icebreakers—*Polar Star* and *Polar Sea*—have exceeded their originally intended 30-year service lives. *Polar Star* was placed in caretaker status on July 1, 2006. Congress in FY2009 and FY2010 provided funding to repair it and return it to service for an additional 7 to 10 years of service; the repair work was completed and the ship was reactivated on December 14, 2012.

On June 25, 2010, the Coast Guard announced that *Polar Sea* had suffered an unexpected engine casualty; the ship was unavailable for operation after that. The Coast Guard placed *Polar Sea* in commissioned, inactive status on October 14, 2011. Section 222 of the Coast

Guard and Maritime Transportation Act of 2012 (H.R. 2838/P.L. 112-213 of December 20, 2012) prohibits the Coast Guard from removing any part of *Polar Sea* and from transferring, relinquishing ownership of, dismantling, or recycling the ship until it submits a business case analysis of the options for and costs of reactivating the ship and extending its service life to at least September 30, 2022, so as to maintain U.S. polar icebreaking capabilities and fulfill the Coast Guard's high latitude mission needs, as identified in the Coast Guard's July 2010 High Latitude Study.

The Coast Guard's third polar icebreaker—*Healy*—entered service in 2000. Compared to *Polar Star* and *Polar Sea*, *Healy* has less icebreaking capability (it is considered a medium polar icebreaker), but more capability for supporting scientific research. The ship is used primarily for supporting scientific research in the Arctic.

With the reactivation of *Polar Star*, the operational U.S. polar icebreaking fleet consists of one heavy polar icebreaker (*Polar Star*) and one medium polar icebreaker (*Healy*). The new polar icebreaker for which initial acquisition funding is requested in the FY2013 budget would replace *Polar Star* at about the time Polar Star's 7- to 10-year reactivation period ends.

The Coast Guard's FY2013 budget initiated a new project for the design and construction of a new polar icebreaker. The project received $7.609 million in FY2013 and $2.0 million in FY2014. The Coast Guard's proposed FY2015 budget requests $6 million to continue initial acquisition activities for the ship.

In July 2011, the Coast Guard provided to Congress a study on the Coast Guard's missions and capabilities for operations in high-latitude (i.e., polar) areas. The study, commonly known as the High Latitude Study and dated July 2010 on its cover, concluded the following:

- **The Coast Guard requires three heavy and three medium icebreakers to fulfill its statutory missions.** These icebreakers are necessary to (1) satisfy Arctic winter and transition season demands and (2) provide sufficient capacity to also execute summer missions. Single-crewed icebreakers have sufficient capacity for all current and expected statutory missions. Multiple crewing provides no advantage because the number of icebreakers required is driven by winter and shoulder season requirements. Future use of multiple or augmented crews could provide additional capacity needed to absorb mission growth.
- **The Coast Guard requires six heavy and four medium icebreakers to fulfill its statutory missions and maintain the continuous presence requirements of the Naval Operations Concept.** Consistent with current practice, these icebreakers are single-crewed and homeported in Seattle Washington.
- **Applying crewing and home porting alternatives reduces the overall requirement to four heavy and two medium icebreakers.** This assessment of non-material solutions shows that the reduced number of icebreakers can be achieved by having all vessels operate with multiple crews and two of the heavy icebreakers homeporting in the Southern Hemisphere.

> Leasing was also considered as a nonmaterial solution. While there is no dispute that the Coast Guard's polar icebreaker fleet is in need of recapitalization, the decision to acquire this capability through purchase of new vessels, reconstruction of existing ships, or commercial lease of suitable vessels must be resolved to provide the best value to the taxpayer. The multi-mission nature of the Coast Guard may provide opportunities to conduct some subset of its missions with non government-owned vessels. However, serious consideration must be given to the fact that the inherently governmental missions

> of the Coast Guard must be performed using government-owned and operated vessels. An interpretation of the national policy is needed to determine the resource level that best supports the nation's interests....
>
> The existing icebreaker capacity, two inoperative heavy icebreakers and an operational medium icebreaker, does not represent a viable capability to the federal government. The time needed to augment this capability is on the order of 10 years. At that point, around 2020, the heavy icebreaking capability bridging strategy expires.[205]

At a July 27, 2011, hearing on U.S. economic interests in the Arctic before the Oceans, Atmosphere, Fisheries, and Coast Guard subcommittee of the Senate Commerce, Science, and Transportation Committee, the following exchange occurred:

> SENATOR OLYMPIA J. SNOWE: On the high latitude study, do you agree with—and those—I would like to also hear from you, Admiral Titley, as well, on these requirements in terms of Coast Guard vessels as I understand it, they want to have—I guess, it was a three medium ice breakers. Am in correct in saying that? Three medium ice breakers.
>
> ADMIRAL ROBERT PAPP, COMMANDANT OF THE COAST GUARD: I agree with the mission analysis and as you look at the requirements for the things that we might do up there, if it is in the nation's interest, it identifies a minimum requirement for three heavy ice breakers and three medium ice breakers and then if you want a persistent presence up there, it would require—and also doing things such as breaking out (inaudible) and other responsibilities, then it would take up to a maximum six heavy and four medium.
>
> SNOWE: Right. Do you agree with that?
>
> PAPP: If we were to be charged with carrying out those full responsibilities, yes, ma'am. Those are the numbers that you would need to do it.
>
> SNOWE: Admiral Titley, how would you respond to the high latitude study and has the Navy conducted its own assessment of its capability?
>
> REAR ADMIRAL DAVID TITLEY, OCEANORGRAPHER AND NAVIGATOR OF THE NAVY: Ma'am, we are in the process right now of conducting what we call a capabilities based assessment that will be out in the summer of this year.
>
> We are getting ready to finish that—the Coast Guard has been a key component of the Navy's task force on climate change, literally since day one when the Chief of Naval Operations set this up, that morning, we had the Coast Guard invited as a member of our executive steering committee.
>
> So we have been working very closely with the Coast Guard, with the Department of Homeland Security, and I think Admiral Papp—said it best as far as the specific comments on the high latitude study but we have been working very closely with the Coast Guard.[206]

Potential issues for Congress regarding Coast Guard polar icebreaker modernization include the following:

- the numbers and capabilities of polar icebreakers the Coast Guard will need in the future;
- the disposition of *Polar Sea*;
- whether the new polar icebreaker initiated in the FY2013 budget should be funded with incremental funding (as proposed in the Coast Guard's Five Year Capital

Investment Plan) or full funding in a single year, as normally required under the executive branch's full funding policy;

- whether new polar icebreakers should be funded entirely in the Coast Guard budget, or partly or entirely in some other part of the federal budget, such as the Department of Defense (DOD) budget, the National Science Foundation (NSF) budget, or both;
- whether to provide future icebreaking capability through construction of new ships or service life extensions of existing polar icebreakers; and
- whether future polar icebreakers should be acquired through a traditional acquisition or a leasing arrangement.

Search and Rescue[207]

General

The possibility of increased sea and air traffic through Arctic waters has increased concerns regarding Arctic-area search and rescue capabilities. Given the location of current U.S. Coast Guard operating bases, it could take Coast Guard aircraft several hours, and Coast Guard cutters days or even weeks, to reach a ship or a downed aircraft in distress in Arctic waters. In addition, the harsh climate complicates search and rescue operations in the region. Particular concern has been expressed about cruise ships that may experience problems and need assistance; there have already been incidents of this kind in recent years in waters off Antarctica. A May 7, 2011, press report stated:

> U.S. and Canadian military commanders say they are examining their rescue capabilities in the Arctic as a shrinking ice cap brought about by climate change opens up rich oil and gas reserves and draws more commercial traffic to the top of the globe....
>
> [U.S. Admiral James Winnefeld, commander of the U.S. Northern Command and the North American Aerospace Defense Command, or NORAD] and NORAD'S deputy commander, Canadian Forces Lt. Gen. Marcel Duval, said more ships in the Arctic Ocean - and more jetliners crossing Arctic skies - could mean more emergencies.
>
> "More ships, more chances of accidents," Duval said in a separate interview in his office at NORAD headquarters at Peterson.
>
> Last summer, the Canadian Coast Guard rescued 197 people from the cruise ship Clipper Adventurer, which ran aground inside the Arctic Circle....
>
> Sea traffic is still light, with only about 25 ships a year currently crossing the maritime Arctic boundary between Alaska and the Yukon. But that number is increasing by 10 to 15 percent a year, according to NORAD statistics.
>
> Civilian air traffic over the Arctic is booming. U.S. and Canadian aviation agencies report more than 9,600 civilian flights across the North Pole in 2010, up nearly 21 percent from 2008.[208]

One option for mitigating the risks associated with cruise ship operations in the Arctic would be to require the vessels to sail in pairs, so that if one ship experiences a problem, the other one could quickly come to its aid.

Increasing U.S. Coast Guard search and rescue capabilities for the Arctic could require one or more of the following: enhancing or creating new Coast Guard operating bases in the region; procuring additional Arctic-capable aircraft, cutters, and rescue boats for the Coast Guard; and adding systems to improve Arctic maritime communications, navigation, and domain awareness.[209] It may also entail enhanced forms of cooperation with navies and coast guards of other Arctic countries.

A May 13, 2013, press report states:

> Arctic nations must urgently improve rescue services in the resource-rich region that is opening up fast to shipping, energy and mining companies, Canadian experts said on Monday [May 13]....
>
> "It is crucial that northern nations strengthen response capabilities to shipping-related accidents foreseen in newly-opened northern waters," scientists from the Canadian Munk-Gordon Arctic Security Program wrote in a report.
>
> Faster responses to emergencies for people living in the Arctic, including floods or forest fires, were also urgently need[ed], the report said.[210]

Coast Guard Statements

A January 2011 press report summarized remarks made by the Commandant of the Coast Guard, Admiral Robert Papp, concerning Coast Guard search and rescue capabilities in the Arctic. The article quoted Papp as saying that there is a need for a range of Coast Guard operational capability for the Arctic, and that "In the meantime, he said the service will lean on partnerships with other Arctic nations. However, he warned that the Coast Guard will likely not be able to respond to any crises in the Arctic circle in a timely fashion. He recalled that the Canadian Coast Guard came under fire when it took six days to rescue a cruise ship and oil tanker that both ran aground in its northern waters." The article stated:

> "We wouldn't be able to make it in six days," he said. "It'd probably take us six weeks to get adequate resources up for a similar thing in our waters, so we have to start focusing on this."[211]

A March 2011 press report summarizing remarks made by Admiral Papp during and after a hearing before the Coast Guard and Maritime Transportation subcommittee of the House Transportation and Infrastructure Committee stated:

> Because retreating ice continues to make the passageways in the Arctic Ocean more navigable, the Coast Guard needs to have air stations in the region to conduct helicopter rescue missions, Papp said. He has not chosen specific cities for the air station sites, but he said the Alaskan cities of Barrow and Kotzebue come to mind.
>
> When lawmakers focus on Arctic resources, the United States' shortage of icebreakers usually draws the bulk of attention, but Papp said the Coast Guard is also in dire need of small boats to conduct rescue missions.[212]

At a June 23, 2011, hearing on the Coast Guard's proposed FY2012 budget before the Oceans, Atmosphere, Fisheries, and Coast Guard subcommittee of the Senate Commerce, Science, and Transportation Committee, the following exchange occurred:

> ADMIRAL ROBERT J. PAPP, JR., COMMANDANT OF THE COAST GUARD: I visited Barrow, Kotzebue, and Nome. And, actually, it was a revisit because I had served up there as an ensign (ph) 35 years ago. And so it was good to get back up there and see the changes.

But what has not changed is the infrastructure up there. And I think that we have to have a robust discussion on the infrastructure needed to support what is no doubt going to be an increase in human activity up there off the – the North Coast of Alaska.

Icebreakers I think are important, but they cloud the discussion of the other needs that we have up there. And I think we've focused on icebreakers over the last few years, even though they are important...

SENATOR MARK BEGICH: Right.

PAPP: ... and needed. But, right now, if we were to have to mount a response like we did in the Gulf of Mexico—I sent 3,000 people down for Deepwater Horizon. You know how many hotel rooms are available in Barrow.

BEGICH: That's right.

PAPP: And we—we have no place to put people up there. We have no hangars for aircraft. We have no piers, no Coast Guard boats. So my immediate, pressing concern is as human activity occurs, as you have that ship that goes through that—first of all, we'll assure safety standards because no matter where ships operate in the world, we are involved in their safety standards.

But if an accident happens, how do we respond? And, right now, we've got zero capability to respond in the Arctic right now. And we've got to do better than that.

That—when people ask me what keeps me awake at night—an oil spill, a collision, a ship sinking in the Arctic keeps me awake at night because we have nothing to respond or, if we respond, it's going to take us weeks to get there.[213]

State Department Statement

The State Department states that

> the United States and the seven other Arctic Council (AC) Member States (Canada, Denmark, Finland, Iceland, Norway, Russian Federation, Sweden) adopted a Ministerial Declaration in April 2009 in Tromso, Norway, which established a Task Force with the mandate of developing a Search and Rescue (SAR) agreement for the Arctic. The United States hosted the first meeting of the Arctic SAR Task Force during December 9-11, 2009, in Washington, D.C. The next round of negotiations is scheduled [sic] for February 25-26, 2010, in Moscow, Russian Federation. The Arctic SAR Task Force has been asked to finalize the Arctic SAR agreement such that it can be presented for adoption by the AC at its Ministerial meeting in Spring 2011.
>
> The United States is at the forefront of efforts to promote safety in the Arctic. The Department of State's Office of Ocean and Polar Affairs is coordinating federal interagency interest to negotiate an instrument for the saving of lives at sea and the rescue of survivors after aircraft accidents in the Arctic. The U.S. Coast Guard is a prominent agency participating in this effort. With so few resources available for SAR in the Arctic, developing a regional agreement to set baseline standards for greater international cooperation and coordination will be of great value.[214]

Arctic Council Agreement on Arctic Search and Rescue (May 2011)

On May 12, 2011, representatives from the member states of the Arctic Council, meeting in Nuuk, Greenland, signed an agreement on cooperation on aeronautical and maritime search and rescue in the Arctic. Secretary of State Hillary Rodham Clinton signed for the United States. Key features of the agreement include the following:

- Article 2 states: "The objective of this Agreement is to strengthen aeronautical and maritime search and rescue cooperation and coordination in the Arctic."

- Article 3 and the associated Annex to the agreement essentially divide the Arctic into search and rescue areas within which each party has primary responsibility for conducting search and rescue operations, stating that "the delimitation of search and rescue regions is not related to and shall not prejudice the delimitation of any boundary between States or their sovereignty, sovereign rights or jurisdiction," and that "each Party shall promote the establishment, operation and maintenance of an adequate and effective search and rescue capability within its area."
- Article 4 and the associated Appendix I to the agreement identify the competent authority for each party. For the United States, the competent authority is the Coast Guard.
- Article 5 and the associated Appendix II to the agreement identify the agencies responsible for aeronautical and maritime search and rescue for each party. For the United States, those agencies are the Coast Guard and the Department of Defense.
- Article 6 and the associated Appendix III to the agreement identify the aeronautical and/or maritime rescue coordination centers (RCCs) for each party. For the United States, the RCCs are Joint Rescue Coordination Center Juneau (JRCC Juneau) and Aviation Rescue Coordination Center Elmendorf (ARCC Elmendorf).
- Article 7 states that "the Parties shall conduct aeronautical and maritime search and rescue operations pursuant to this Agreement consistent" with certain general guidelines.215
- Article 8 states that "a Party requesting permission to enter the territory of a Party or Parties for search and rescue purposes, including for refueling, shall send its request to a search and rescue agency and/or RCC of the relevant Party or Parties," and that "the Party receiving such a request shall immediately confirm such receipt. The receiving Party, through its RCCs, shall advise as soon as possible as to whether entry into its territory has been permitted and the conditions, if any, under which the mission may be undertaken."
- Article 9 states that "the Parties shall enhance cooperation among themselves in matters relevant to this Agreement," that "the Parties shall exchange information that may serve to improve the effectiveness of search and rescue operations," and that "the Parties shall promote mutual search and rescue cooperation by giving due consideration to collaborative efforts."
- Article 10 states that "the Parties shall meet on a regular basis in order to consider and resolve issues regarding practical cooperation."
- Article 11 states that "after a major joint search and rescue operation, the search and rescue agencies of the Parties may conduct a joint review of the operation led by the Party that coordinated the operation."
- Article 12 states that "unless otherwise agreed, each Party shall bear its own costs deriving from its implementation of this Agreement," and that "implementation of this Agreement shall be subject to the availability of relevant resources."
- Article 18 states that "any Party to this Agreement may, where appropriate, seek cooperation with States not party to this Agreement that may be able to contribute to the conduct of search and rescue operations, consistent with existing international agreements."

- Article 19 states that "any Party may at any time withdraw from this Agreement by sending written notification thereof to the depositary216 through diplomatic channels at least six months in advance."[217]

Figure 4 shows an illustrative map of the national areas of search and rescue responsibility based on the geographic coordinates listed in the Annex to the agreement.

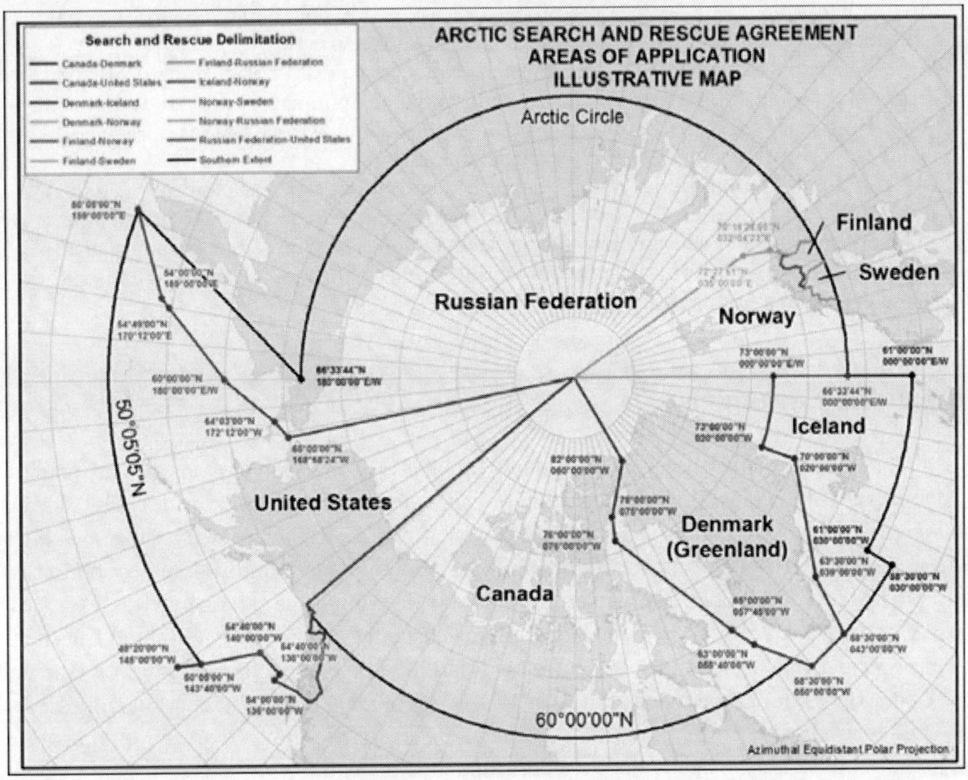

Source: "Arctic Search and Rescue Agreement," accessed July 7, 2011, at http://www.arcticportal.org/features/ features-of-2011/arctic-search-and-rescue-agreement.

Figure 4. Illustrative Map of Arctic SAR Areas in Arctic SAR Agreement. (Based on geographic coordinates listed in the agreement).

The State Department states that the agreement:

> is the first legally-binding instrument negotiated under the auspices of the Arctic Council. It coordinates life-saving international maritime and aeronautical SAR coverage and response among the Arctic States across an area of about 13 million square miles in the Arctic.
>
> As Arctic sea ice coverage decreases, ship-borne activities are increasing significantly in the Arctic. Flight traffic is also on the rise as new polar aviation routes cross the Arctic air space in several directions. As human presence and activities in the Arctic expand, the potential for accidents increases as well. Limited rescue resources, challenging weather conditions, and the remoteness of the area render SAR operations difficult in the Arctic, making coordination among the Arctic nations imperative. The

SAR Agreement will improve search and rescue response in the Arctic by committing all Parties to coordinate appropriate assistance to those in distress and to cooperate with each other in undertaking SAR operations. For each Party, the Agreement defines an area of the Arctic in which it will have lead responsibility in organizing responses to SAR incidents, both large and small. Parties to the Agreement commit to provide SAR assistance regardless of the nationality or status of persons who may need it.

The Arctic Council launched this initiative at its 2009 Ministerial Meeting in Tromso, Norway, establishing a Task Force, co-chaired by the United States and the Russian Federation. The Task Force proceeded in a highly collaborative spirit, meeting five times (in Washington, Moscow, Oslo, Helsinki and Reykjavik).

The signature of the SAR Agreement in Nuuk is a positive step toward building partnerships in the Arctic. In particular, it reflects the commitment of the Arctic Council States to enhance their cooperation and offer responsible assistance to those involved in accidents in one of the harshest environments on Earth.

This Agreement illustrates one of the most successful negotiations to date to address emerging issues in the Arctic. Arctic Council participants approached SAR negotiations with collaboration and dedication to a positive outcome. The United States congratulates its colleagues in this effort and looks forward to further collaboration on the vital issues facing the rich but fragile Arctic region.[218]

Geopolitical Environment[219]

In recent years, many observers have noted that the loss of Arctic ice[220] is already leading to stepped-up human activity in the high north, particularly in the form of increasing commercial traffic and economic development. This trend has brought forth a range of issues on the geopolitical front, from environmental protection to search-and-rescue capabilities to the delineation of national boundaries—which will determine access to natural resources. These concerns are being addressed cooperatively in both bilateral and multilateral fashion, especially under the aegis of the Arctic Council and the U.N. Convention on the Law of the Sea (UNCLOS). International law Professor William Moomaw of the Fletcher School of Law and Diplomacy has noted that "the lure of riches in the Arctic draws ever more companies and nations, but so far it's been relatively amicable jousting and jostling there."[221]

Nonetheless, some observers continue to raise questions concerning security in the high north, and have advanced sometimes conflicting views regarding the potential roles of military forces in the region. Of the other Arctic coastal nations, the United States enjoys strong political and economic ties with Canada, Norway, and Denmark; all four countries are members of NATO. Although the United States views Russia as an important partner in developing policies to cope with changing conditions in the Arctic, relations with Moscow have had numerous areas of tension in recent years. Several non-Arctic nations, including India and China, have also evinced interest in the high north, and recently gained permanent observer status in the Arctic Council. In addition, the European Union, which sought but was denied full observer status in 2009 and 2013, is nonetheless developing policy toward the Arctic.[222]

Multilateral Political Cooperation

As noted elsewhere in this report, in May 2008, ministerial representatives of the five Arctic littoral states attended a meeting convened by Denmark in Ilulissat, Greenland (a semiautonomous territory of Denmark). Danish Foreign Minister Stig Møller implied that the meeting was intended to develop interim measures for Arctic governance: "We must continue to fulfill our obligations in the Arctic area until the U.N. decides who will have the right to the sea and the resources in the region. We must agree on the rules and what to do if climate changes make more shipping possible."[223] Attendees discussed a variety of issues, including the environment, transportation, resources, and security. The Danish Foreign Ministry stated that the resulting Ilulissat Declaration

> sent a clear political signal to the local inhabitants and the rest of the world that we will act responsibly when addressing the development in the Arctic Ocean. We have committed ourselves politically to solve any disagreements through negotiation. Thus, hopefully, we have eradicated all the myths about a 'race for the North Pole.' The legal framework is in place and the five States have now declared that they will abide by it.[224]

The Arctic Council

The Arctic has increasingly become a subject of discussions in bilateral meetings among leaders of the nations in the region and elsewhere around the globe. The main international forum for cooperation in the high north, however, is the eight-nation Arctic Council, of which the United States is an active member.

A series of meetings initiated by Finland in 1989 eventually led to the creation of the Arctic Council in 1996. The Council has become the primary intergovernmental "high level forum" for cooperation in the Arctic region. It addresses a wide range of issues, including regional development, the environment, emergency response, climate change, and natural resource extraction.

The Council membership consists of the eight countries that have sovereign territory within the Arctic Circle: the United States, Canada, Norway, Denmark (by virtue of its territory Greenland), Russia, Sweden, Finland, and Iceland. Only these countries have voting rights. Six indigenous Arctic peoples' organizations are permanent participants. Permanent observer status is held by France, Germany, the Netherlands, Poland, Spain, the UK, China, India, Italy, Japan, South Korea, and Singapore; the latter six were added during the May 2013 summit meeting. Also represented on the Council are several intergovernmental and nongovernmental observers, including the International Red Cross, the United Nations Development Program, the Nordic Council, and the Worldwide Fund for Nature.

The Council has six working groups devoted to various issues: (1) the Arctic Contaminants Action Program; (2) the Arctic Monitoring and Assessment Program; (3) Conservation of Arctic Flora and Fauna; (4) Emergency Prevention, Preparedness and Response; (5) Protection of the Arctic Marine Environment; and (6) the Sustainable Development Working Group. The United States reportedly vetoed security as an issue of consideration for the Council.[225]

Each member state is represented by a Senior Arctic Official (SAO), who is usually drawn from that country's foreign ministry. The SAOs hold meetings every six months. The Council convenes ministerial-level meetings every two years, at the end of each chairmanship, while the working groups meet more frequently. The Council has a two-year rotating chairmanship. In May 2013, Sweden passed the gavel to Canada. The United States will take over in 2015. The United States last held the chairmanship from 1998 to 2000.

In May 2011, then-U.S. Secretary of State Hillary Clinton attended the Arctic Council ministerial summit, held in Nuuk, Greenland; she was accompanied by then-U.S. Interior Secretary Ken Salazar. They were the first U.S. Cabinet members to attend an Arctic Council meeting, and observers stated that their visits served to raise the profile of Arctic issues. Noting the increased commercial activity in the region, Secretary Clinton declared, "We need to pursue these opportunities in a smart, sustainable way that preserves the Arctic environment and ecosystem."[226] Among other issues, attendees focused on efforts to reduce emissions that cause "black carbon" to settle on the Arctic region, accelerating ice melt. In addition, the Council discussed launching a longer-term study on methods to cope with possible future oil spills.

The major "deliverable" of the summit, however, was the signing of an Agreement on Cooperation on Aeronautical and Maritime Search and Rescue (SAR) in the Arctic. The accord was the "first legally-binding instrument negotiated under the auspices of the Arctic Council."[227] The SAR initiative, developed mainly by the United States and Russia, had been introduced during the April 2009 summit. The Council also approved the establishment of a permanent secretariat, to be based in Tromsø, Norway.

In June 2012, then-Secretary Clinton traveled once more to the Arctic, visiting the newly opened secretariat in Tromsø as part of an eight-day trip to Scandinavia. While there, she emphasized that the U.S. government "want[s] the Arctic Council to remain the premier institution that deals with Arctic questions."[228]

The Council held its most recent summit on May 15, 2013 in Kiruna, Sweden. Secretary of State John Kerry's attendance, following on the heels of Secretary Clinton's two years earlier, underlined the growing importance attached by the U.S. government to polar issues. The Council revisited several of its core topics, including ocean acidification, ice loss, black carbon, biodiversity, the status of indigenous peoples, and emergency preparedness. Consideration of the latter issue resulted in the approval of a second Council accord: the Agreement on Cooperation on Marine Oil Pollution Preparedness and Response in the Arctic.

From a geopolitical perspective, perhaps the most important step taken by the Council at the May 2013 summit was enlargement; six countries—China, India, Singapore, South Korea, Japan, and Italy—were approved for permanent observer status.[229] Prior to the meeting, the five Nordic countries reportedly had endorsed the admission of new observers, while Russia and Canada were said to be opposed—the latter for fear of diluting the interest of indigenous peoples.[230] Washington reportedly played its cards close to the vest, remaining silent on the issue until the day of the decision. Secretary of State John Kerry is said to have brokered the compromise. Not long before the conference, one analyst summarized the arguments for and against (mostly for) adding new seats to the table:

> Openness would be a shrewd move, at a stroke enhancing the council's legitimacy and the quality of its deliberations, and reducing the risk of being bypassed by countries acting unilaterally. Some of the council's recent applicants could also represent a financial boon: China, South Korea, India and Japan have the fastest-growing stable of Arctic scientists, while the EU last year proposed devoting €80 billion to Arctic research. Meanwhile, the risks of admitting new members, even heavy-hitters like China, are low. First, observers' powers are limited: they may engage with the forum's working groups, propose projects and state their views (all with permission), but they have no vote. Secondly, observer status does not alter international law: under UNCLOS [United Nations Convention on the Law of the Sea], extra-regional actors have no jurisdiction in

> Arctic waters, and no applicant disputes this. Thirdly, the council is still fundamentally devoted to promoting research and knowledge sharing, and new observers could contribute greatly in this regard. Some Permanent Participants also fear being marginalised if new observers are admitted, but the same countries could still engage bilaterally with Arctic states from outside of the council, and there is no difference in power between permanent and ad hoc observer status. The main danger of admitting too many new observers is therefore likely to be confined to a diminished capacity for reaching swift consensus.[231]

According to some observers, the decision to add India and Singapore—which are situated thousands of miles from the Arctic Circle—would seem to indicate that the Council is transforming itself from a regional to an international forum. As noted, the revolving chair of the Council now passes from Scandinavia to North America—Canada will hold the post until 2015, when the United States will take over. Patrick Borbey, Chair of Canadian Senior Arctic Officials, stated that his government hopes the Council will "evolve from a very solid organization doing great scientific assessment to one that actually makes policy and implements it ... and makes sure it's monitored on an ongoing basis."[232] Canada's stated priorities for its chairmanship are intended to benefit the 4 million people of the north through responsible resource development, safe shipping, and the promotion of sustainable circumpolar communities. Canada also will seek to strengthen the Arctic Council in an effort to "enhance the capacity of the Permanent Participant organizations, improve the Council's coordination and maximize efficiencies."[233]

Russia

It has been noted that Russia "has at least half of the Arctic in terms of area, coastline, population and probably mineral wealth."[234] Moscow is keen to capitalize on natural resource development and shipping in the region. As noted elsewhere in this report, Russia and Norway in late 2010 resolved a 40-year dispute over national borders in the Barents Sea; the accord permits exploration for undersea oil, believed to be in rich supply there. In March 2013, it was announced that Russia and China had signed an agreement under which China would purchase oil from Russia in exchange for exploration licenses in the Arctic. In addition, scientists estimate that the sea route along the Siberian coast (referred to as the Northeast Passage or the Northern Sea Route, or NSR) will be ice-free and navigable well before the Northwest Passage through the Canadian archipelago.[235] Indeed, the route is already being used: 34 vessels traversed the NSR in 2011, 46 sailed the passage in 2012, and 71 did so in 2013.[236] Russia sees significant economic opportunities in offering icebreaker escorts, refueling posts, and supplies to the commercial ships that will ply the waterway.[237] The NSR will cut transport times and costs for Russian ships as well—the development of the shale gas industry in the United States means that Russia will likely now turn to Asia to market its liquefied natural gas (LNG).[238] But analysts note that Russia's ability to capitalize on new opportunities will require international cooperation and goodwill.

A demonstration of such cooperation took place in August 2012, when Russia joined the United States and Norway in the Barents Sea for the biannual *Northern Eagle 2012* naval maneuvers, an exercise that Norwegian Defense Minister Espen Barth Eide referred to as "a strong and interest-based neighborhood cooperation with Russia."[239] However, planning for the 2014 exercise, which was scheduled to be held in May, was put on hold. In the wake of Russia's annexation of Crimea and its efforts to further destabilize Ukraine, some analysts have questioned whether Moscow will continue to pursue a policy of multilateral cooperation in the Arctic. Other Arctic Council member states have begun to push back. In protest of Russia's recent actions in Crimea/Ukraine, Canada announced that it would not participate in an April 2014 working-level-group Arctic Council meeting in Moscow. In addition, former Secretary of State Hillary Clinton, during whose tenure a "reset" in relations with Russia was sought, reportedly stated that Arctic cooperation may be jeopardized if Russia pursues expansionist policies in the High North.[240]

China

China expert Elizabeth Economy has noted that Beijing "has begun the process of engaging in the Arctic through research, investment, and diplomacy."[241] Although its borders lie some distance from the Arctic, China has displayed a growing interest in the region, based mainly upon the potential opportunities for shorter sea routes and the eventual development of energy-related natural resources, as well as metals and minerals; the International Institute of Strategic Studies states that "China is the world's largest consumer of raw materials and hydrocarbon resources.[242] China's economy is strongly dependent upon exports; some analysts have estimated that as much as one-half of China's GDP is reliant upon exports and shipping. Ocean transportation is the chief avenue for China's large petroleum imports. Beijing is keenly interested in having free access to the future northern waterways, which would drastically reduce both sailing times and transportation costs. Other observers have argued that China's interest is motivated in the Arctic's emerging status as "the new fishing grounds—the world's largest storehouse of biological protein."[243] Finally, some have also interpreted China's growing interest in the Arctic as further evidence of its effort to expand its influence as a global player: "They know that [the] Arctic may be one of the hot spots of the 21st century."[244]

Many Chinese believe that the Arctic should be considered as part of the "global commons." According to David Curtis Wright of the U.S. Naval War College, "The mantra that the Arctic and its natural resource wealth belong to no one country but constitute the common heritage of all humankind is virtually *de rigueur* in recent Chinese public commentary on Arctic affairs."[245] This attitude was reflected during a May 2012 workshop on Sino-Nordic cooperation hosted in Beijing, when Chinese participants referred to their country as a "near-Arctic state" and a "stakeholder." Some analysts believe that China will likely remain officially circumspect on this question, as its "foreign policy rests on a profound respect for territorial integrity."[246]

Nonetheless, some Chinese analysts reportedly are encouraging their government to challenge Canada's claim of sovereignty over the Northwest Passage.

In recent years, China has been cultivating relationships with the Nordic countries. In April 2012, former Chinese Premier Wen Jiabao visited Sweden and Iceland, and two months later former President Hu Jintao went to Denmark; the two leaders reportedly discussed large-scale investments in the region, and in February 2014, a Greenlandic representative was in Beijing to discuss economic cooperation. In April 2013, China and Iceland signed a free trade agreement— China's first such pact with a European government. In addition, China (like several other nations) has established a research station in the Svalbard archipelago and has beefed up the size of its embassy staff in Iceland. In October 2013, Chinese Vice Premier Ma Kai met with Icelandic President Grimsson and stated that "China is willing to expand pragmatic [cooperation] with Iceland in fields of economy, trade, geothermal energy, Arctic research environment, science and technology and social development."[247]

China has been active in conducting research on the Arctic; in 1994, Beijing purchased from Ukraine a research icebreaker it named the *Xuelong,* and has constructed a state-of-the-art polar capable research vessel, the *Snow Dragon*; in August 2012, the latter conducted a trans-Arctic voyage, from Shanghai to Iceland; it was China's fifth Arctic research expedition. In September 2013, the *Yong Shen, a* Chinese cargo ship, became the first commercial vessel to complete the voyage from Asia to Rotterdam via the Northern Sea route.[248]

United States

The attendance by Secretaries of State Clinton and Kerry of the last two Arctic Council summits, as mentioned above, indicates that the Obama Administration has placed some degree of priority on the far north. Consistent with this, the Obama Administration has recently updated national policy toward the region.

As noted elsewhere in this report, the Bush Administration in early January 2009 issued a presidential directive outlining its policy on the Arctic region; the last such directive had been issued in 1994. The Obama Administration operates under the Bush Administration's policy directive and augmented it on May 10, 2013, when it announced its National Strategy for the Arctic Region. The new policy blueprint identifies three major "lines of effort": (1) Advancing United States security interests; (2) Pursuing responsible Arctic region stewardship; and (3) Strengthening international cooperation. These activities are to be guided by four principles: (1) Safeguard peace and stability; (2) Make decisions using the best available information; (3) Pursue innovative arrangements among various levels of government and the private sector; and (4) Consult and coordinate with Alaska natives.

Regarding the Arctic Council, the National Strategy states that "[w]orking through bilateral relationships and multilateral bodies, including the Arctic Council, we will pursue arrangements that advance collective interests, promote shared Arctic state prosperity, protect the Arctic environment, and enhance regional security, and we will work toward U.S. accession to the United Nations Convention on the Law of the Sea (Law of the Sea Convention)." Thus, the Council is regarded as one of the organizations, though not the only one, through which the United States will seek to fulfill its goals for the region. The Strategy highlights the Arctic Council under its third "line of effort"—that of strengthening international cooperation.

> In recent years, the Arctic Council has facilitated notable achievements in the promotion of cooperation, coordination, and interaction among Arctic states and Arctic indigenous peoples. Recent successes of the Council include its advancement of public safety and environmental protection issues, as evidenced by the 2011 Arctic Search-and-Rescue Agreement and by the 2013 Arctic Marine Oil Pollution Preparedness and Response Agreement. The United States will continue to emphasize the Arctic Council as a forum for facilitating Arctic states' cooperation on myriad issues of mutual interest within its current mandate.[249]

Besides the Arctic Council, the strategy document appears to suggest that the Administration will adopt a multilateral approach to problem-solving, utilizing unspecified "existing partnerships," "multilateral fora," "new arrangements," and "efficient and effective joint ventures" to address emerging challenges in the high north.

The strategy defines the issue of security in the widest sense: "U.S. security in the Arctic encompasses a broad spectrum of activities, ranging from those supporting safe commercial and scientific operations to national defense." Some of these issues involve solutions that can be viewed as appropriate for the Arctic Council to generate, such as the above-mentioned 2011 search-and-rescue agreement. On the issue of national defense, which concerns protection of state sovereignty, many of the issues, such as developing infrastructure and enhancing domain awareness, are primarily domestic in nature. Concerning freedom of navigation, the strategy document states that the United States will rely upon "[e]xisting international law [which] provides a comprehensive set of rules governing the rights, freedoms, and uses of the world's oceans and airspace, including the Arctic." The document states: "we encourage Arctic and non-Arctic states to work collaboratively through appropriate fora to address the emerging challenges and opportunities in the Arctic region, while we remain vigilant to protect the security interests of the United States and our allies."

The document also frequently asserts that the interests of the Arctic region's indigenous peoples must be protected. As noted above, indigenous Arctic peoples' organizations have permanent participant status on the Arctic Council. According to the Council's website, they "have full consultation rights in connection with the Council's negotiations and decisions." The Council thus represents a potentially important venue for the Administration's goal of ensuring that their voices are heard and heeded.

In a speech at the aforementioned Arctic Council's ministerial in May 2013, Secretary of State Kerry stated that there was considerable overlap in interests among the eight member states, and that their decisions "don't stop at the 66th parallel." He cited climate change as the most far-reaching issue, and also noted other areas of concern, including "acidification, pollution, ice melt, rising sea levels, disappearing species, and indiscriminate development practices" that can have an effect "downstream" on further challenges "to our economies, to our national security, and to international stability." He lauded the work of the Arctic Council, "which addresses these challenges."[250]

Finally, the U.S. Department of State has raised the issue of fisheries, stating that as stocks move northward, there will be a need to adapt current fisheries management mechanisms and create new ones for regions not currently being fished. According to the State Department, "The United States is ... considering whether it would be desirable for a group of States with interests in present and future Arctic fisheries to adopt some form of general statement or declaration." While the Arctic Council's Kiruna Declaration[251] does not specifically mention fisheries management, the Council's goal of promoting sustainable and responsible use of natural resources make it a potential venue for the Administration to explore interest in adopting a general statement on the issue.

At the end of January 2014, the White House released the Implementation Plan for the National Strategy for the Arctic Region. The plan essentially follows and complements the objectives spelled out in the May 2013 Strategy document. Under the rubric of security, for example, the Implementation Plan calls for the development of communication infrastructure to support aerial and maritime transportation, and to improve domain awareness. Under the heading of Arctic stewardship, the plan lists a wide array of topics, including developing greater understanding of environmental and ecological processes, and improving conditions for indigenous peoples through addressing issues concerning health, cultural heritage, and community sustainability. On the international level, the plan calls for greater preparedness on oil pollution cleanup and searchand-rescue capabilities, as well as safeguarding marine ecosystems. It calls for the development of a "robust agenda" for the upcoming U.S. chairmanship of the Arctic Council, and for U.S. accession to the Law of the Sea Convention.[252] On February 13, 2014, Alaska Senator Lisa Murkowski stated in a letter to the President that she was "severely disappointed" with the implementation plan, which, she argued, "provides a snapshot of existing Arctic-related programs and projects with numerous assessments to be undertaken, but no real path of action." The following day, Secretary of State Kerry announced the creation of a new, high-level position: Special Representative for the Arctic Region. In a press statement, Secretary Kerry stated that he would "look forward to work closely with Alaska's Congressional delegation to strengthen American's engagement in Arctic issues."[253]

Security Issues

Throughout the Cold War, the Arctic region was a zone of major strategic interest, where the United States, the Soviet Union, and allied states conducted air and naval maneuvers and tested ballistic missiles. With the collapse of the Warsaw Pact and the Soviet Union, however, the importance of the high north diminished in the 1990s. Although the establishment of the Arctic coastal states' sovereignty through the demarcation of boundaries in the region is being conducted peacefully under the auspices of the U.N. Convention on the Law of the Sea, the Arctic is once again being viewed by some as a potential emerging security issue. In a December 2011 *Washington Post* op-ed, Heather Conley, a senior fellow at the Center for Strategic and International Studies, noted several developments:

> In April [2011], President Obama signed a new command plan that gives NORAD and the U.S. Northern Command greater responsibility in protecting the North Pole and U.S. Arctic territory.... In 2009, Norway moved its operational command to its northern territories above the Arctic Circle. Russia has plans to establish a brigade that is specially equipped and prepared for military warfare in Arctic conditions. Denmark has made it a strategic priority to form an Arctic Command. Canada is set to revitalize its Arctic fleet, including spending $33 billion to build 28 vessels over the next 30 years.[254]

Similarly, Canadian academic Rob Huebert pointed out that in August 2010 the United States, Canada, and Denmark conducted in the Canadian Arctic their annual joint naval exercises involving several advanced and powerful warships. Huebert observed that "while defence officials are quick to point out they see no military threat to the region, it's still interesting to see these three Arctic friends coming together to improve their naval combat capability in the Far North."[255] In varying degrees, the Arctic coastal states have indicated a willingness to establish and maintain a military presence in the high north.[256]

Although some have argued that terrorism and hijacking may constitute security concerns in the region, others maintain that such threats are chimerical, given the challenges of distance and geography, and the difficulty of navigating in a polar environment. *The Economist* has asserted that "the risks of Arctic conflict have been exaggerated. Most of the Arctic is clearly assigned to individual countries. According to a Danish estimate, 95% of Arctic mineral resources are within agreed national boundaries."[257] Other factors may also postpone energy exploration. For example, in the *New York Times*, three scholars noted that "the shale gas revolution is already delaying some Arctic energy projects." In addition, some companies are reportedly "fearful of the financial and public relations risk of working in the pristine icy wilderness."[258]

A report by the Arctic Institute noted that "[t]he armed forces, beyond their responsibility for handling all contingencies, are also the only agencies with both the requisite monitoring instruments and the physical capabilities to operate in such a vast and inhospitable region."[259] However, as mentioned above, the Arctic Council does not address regional security issues. To fill this apparent void, a report by the Washington, DC-based Center for Strategic and International Studies has proposed the creation of a separate organization, the Arctic Coast Guard Forum (ACGF), consisting initially of the eight Arctic Council states, but possibly expanding eventually to include other countries willing to contribute assets. The ACGF, which could potentially be headquartered at the U.S. Air Force base in Thule, Greenland, would "focus first on information sharing yet should also seek to develop methods of cooperation in support of the Arctic Council's search-and-rescue agreement and future international oil spill response agreement."[260]

However, other, relatively little-publicized multilateral discussions of security issues have already been taking place. In mid-2011, the U.S. European Command (EUCOM), in cooperation with the Norwegian Ministry of Defense, established the Arctic Security Forces Roundtable (ASFR), consisting of high-ranking military officers from the eight members of the Arctic Council, plus France, Germany, the Netherlands, and the UK. Their first meeting, held in June 2011 in Oslo, addressed a range of issues, including infrastructure, the environment, joint exercises and training, and marine domain awareness. In August 2012, the ASFR held a conference Bodo, Norway; the meeting focused mainly on how to improve the communications infrastructure in the high north.[261] Another newly formed venue at which military leaders discuss Arctic issues is the Northern Chiefs of Defense conference, the first of which was held in Goose Bay, Labrador, in May 2012; it was attended by military representatives from the eight Arctic Council governments.[262]

NATO

The Arctic has also become a region of interest for NATO. However, as one writer has noted, "[t]here is currently no consensus within the alliance that NATO has any role to play in the Arctic, as Canada strongly opposes any NATO involvement on sovereignty grounds and other NATO members are concerned with negative Russian reaction."[263] Speaking in Reykjavík in January 2009, former NATO Secretary General Jaap de Hoop Scheffer urged that member states not allow the Arctic to become a divisive issue. He also recommended that the alliance and Russia cooperate through building upon their shared experience in search-and-rescue operations. Former Danish Prime Minister Anders Fogh Rasmussen, who became secretary general of the alliance in August 2009, has also addressed security in the high north. Citing the "potentially huge security implications" of Arctic climate change, Rasmussen in October 2009 stated that "I think it is within the natural scope of work for NATO to be the forum for consultation and discussion on [selected Arctic] issues." In March 2009, however, Russia's NATO ambassador stated that Moscow would not cooperate with the alliance on Arctic matters. And in September 2010, then-Russian President Medvedev reportedly observed that "the Arctic can do fine without NATO," and that his government "views [possible NATO] activity with quite serious tension, because it is after all a zone of peaceful cooperation, economic cooperation, and of course the military factor always—at a minimum—creates additional questions." On a visit to Moscow in November 2010,

Rasmussen assured the Russians that NATO does not intend to establish a presence in the Arctic.[264] Since 2006, several member and partner states have participated in *Cold Response*, a wide-ranging annual crisis response exercise hosted by Norway. During the most recent joint maneuvers, held in March 2014, 16 nations fielded air, land, and naval assets and approximately 16,000 troops. Although the exercises are multilateral, they are not conducted under the auspices of NATO.[265]

A 2013 NATO Parliamentary Assembly report noted that "50% of the territory surrounding the Arctic Sea is a territory of a NATO member state," and suggested that "NATO could serve as a forum for dialogue on military issues...."[266] The report contends that the alliance is well-equipped to play a key role in addressing security challenges that will likely emerge, particularly those that involve surveillance, search-and-rescue, and environmental cleanup. However, observers note that the lack of unanimity over a NATO presence in the Arctic is reflected by the fact that the high north is mentioned neither in the alliance's 2010 strategic concept, nor in the final declaration of the 2012 Chicago summit. On May 8, 2013, following a visit to Norway, Secretary General Rasmussen stated that "at the present time," the alliance had "no intention of raising its presence and activities in the High North." He later twittered that "the Arctic is a harsh environment. It rewards cooperation, not confrontation. I trust we'll continue to see cooperation."[267]

Writing in an Atlantic Council blog, Dr. Page Wilson, of the UK's Royal Military Academy Sandhurst, has argued that, for the present, "NATO's reluctance to increase its focus on the Arctic may appear curious," whereas, "[i]n fact, it is eminently sensible." She notes that the alliance is already active in the region, with regular military exercises (such as the above-mentioned *Cold Response*), the air policing mission over Iceland, and the presence in Greenland of elements of the U.S. Active Layered Theater Ballistic Missile Defense. In addition, she notes that several international fora, chief among them the Arctic Council, are fostering cooperation in areas such as environmental, social, and economic issues, and that discussion of security matters can be deferred "until such time as greater clarity and agreement emerges about the nature of the Arctic as a political space." However, another scholar has noted that the March 28, 2014, appointment of former Norwegian Prime Minister Jens Stoltenberg as the alliance's new Secretary-General "should alert global attention to ... the Arctic."[268]

Russia

The Russian government has stated that, although it deplores the notion of an arms race in the high north and does not foresee a conflict there, it intends to protect its Arctic interests.[269] However, Russia has at times appeared to be sending out mixed messages in this regard. For example, at the conclusion of a meeting in September 2010, Russian Foreign Minister Sergei Lavrov and former Canadian Foreign Minister Lawrence Cannon stated that "any militarization [of the Arctic] is out of the question." And in June 2011, then-Prime Minister (and currently President) Vladimir Putin stated, "Russia will definitely expand its presence in the Arctic. We are open for dialogue with our foreign partners and with all neighbors in the Arctic region. But we will naturally defend our own geopolitical interests firmly and consistently." The following month, Putin announced plans to build a large shipping port on the Yamal peninsula, and the government stated that it would be sending two brigades to the north to protect its interests. In addition, Russia is reportedly rebuilding former Soviet airbases in the region. In addition, President Putin in December 2013 stated that there was a need for "every lever for the protection of its security and national interests" in the Arctic, and ordered the development by 2014 of a strategic command in the region. However, as noted above, in order to reap the economic benefits of natural resources development and shipping, the Russians will need to rely heavily on foreign capital and technology, and, according to Canadian Arctic specialist Michael Byers, "probably they realize how expensive it would to take another approach [than cooperation], especially one involving militarization."[270]

However, some analysts believe that Russia's recent occupation of Crimea and its continuing interference in Eastern Ukraine may have repercussions for Arctic cooperation in the security arena. In March 2014, Norway announced that it was suspending scheduled military activities with Russia.[271] On April 1, 2014, NATO Foreign Ministers issued a statement condemning Russia's intervention in Ukraine and refused to recognize Moscow's "illegal and illegitimate attempt to annex Crimea," and expressed "grave concern over the authorisation by the Russian Parliament to use the armed forces of the Russian Federation on the territory of Ukraine." Later in April, the defense ministers of the five Nordic states met to discuss military cooperation in the Arctic. In addition, Alaska Senator Mark Begich recently stated that "with unpleasant reminders of the Cold War, ... a strong military presence in the Arctic is more important than ever."[272]

China

Some Chinese leaders also have voiced concern over perceived emerging security issues in the Arctic. In early March 2010, a Chinese admiral stated that "the current scramble for the sovereignty of the Arctic among some nations has encroached on many other countries' interests," and added that China had to "make short and long term ocean strategic development plans to exploit the Arctic because it will become a future mission for the navy." Some analysts, however, believe that China's general approach toward the Arctic will remain decidedly low-key: "To date, China has adopted a wait-and-see approach to Arctic developments, wary that active overtures would cause alarm in other countries due to China's size and status as a rising global power." China is believed to be keen on resolving through diplomacy the national interests of both littoral and non-Arctic states in the high north. Toward that end, it sought permanent observer status on the Arctic Council. Its candidacy reportedly was delayed by a dispute with Norway, which in 2010 awarded the Nobel Peace Prize to Chinese dissident Liu Xiaobo.[273] It was reported in early April 2013 that Norway would support China's bid for observer status at the May Arctic Council meeting. As noted above, China was approved at the meeting for observer status.[274]

U.S. Military Forces and Operations[275]

During the Cold War, the Arctic was an arena of military competition between the United States and the Soviet Union, with both countries, for example, operating nuclear-powered submarines, long-range bombers, and tactical aircraft in the region. The end of the Cold War and the collapse of most elements of the Russian military establishment following the dissolution of the Soviet Union in December 1991 greatly reduced this competition and led to a reduced emphasis on the Arctic in U.S. military planning.

The diminishment of Arctic sea ice is now leading U.S. military forces to pay renewed attention to the Arctic. This is particularly true in the case of the Navy and Coast Guard, for whom diminishment of Arctic sea ice is opening up potential new operating areas for their surface ships. Navy and Coast Guard activities relating to the Arctic are taking place as other countries, such as Russia, Canada, and Norway, are examining the potential implications for their military forces of diminished Arctic sea ice, and taking or contemplating steps to increase their own navy and coast guard presence and operations in the region.[276] The U.S. Army, too, is beginning to focus more on Arctic operations.[277]

Issues such as sovereignty, freedom of navigation, and energy exploration have the potential for leading to either cooperation or competition in the Arctic, depending on how they are handled. U.S. military officials and military officials from other Arctic states often stress the cooperative aspects of the situation and suggest that the competitive aspects of the situation have been exaggerated in some press accounts. Renewed tensions with Russia following Russia's annexation of Crimea, however, have led some observers to express concern regarding the potential for renewed military tension and competition in the Arctic.[278]

DOD in General

2010 Quadrennial Defense Review (QDR)

The Department of Defense's (DOD's) report on the 2010 Quadrennial Defense Review (QDR), submitted to Congress in February 2010, states:

> The effect of changing climate on the Department's operating environment is evident in the maritime commons of the Arctic. The opening of the Arctic waters in the decades ahead[,] which will permit seasonal commerce and transit[,] presents a unique opportunity to work collaboratively in multilateral forums to promote a balanced approach to improving human and environmental security in the region. In that effort, DoD must work with the Coast Guard and the Department of Homeland Security to address gaps in Arctic communications, domain awareness, search and rescue, and environmental observation and forecasting capabilities to support both current and future planning and operations. To support cooperative engagement in the Arctic, DoD strongly supports accession to the United Nations Convention on the Law of the Sea.[279]

The report also states:

> The Department of Defense and its interagency partners must be able to more comprehensively monitor the air, land, maritime, space, and cyber domains for potential direct threats to the United States. Such monitoring provides the U.S. homeland with an extended, layered in depth defense. This effort includes enhanced coordination with Canada for the defense of North America as well as assisting Mexico and Caribbean partners in developing air and maritime domain awareness capacities. Special attention is required to develop domain awareness tools for the Arctic approaches as well. In coordination with domestic and international partners, DoD will explore technologies that have the potential to detect, track, and identify threats in these spheres to ensure that capabilities can be deployed to counter them in a timely fashion.[280]

The report further states:

> Central to the security of the United States is a strong transatlantic partnership, which is underpinned by the bilateral relationships between the United States and the governments of Europe. We will continue to work with this community of like-minded nations, whether by engaging with allies still shaping their democracies after decades of living in the shadow of the Soviet Union, building on the benefits of French reintegration into NATO's military structure, or addressing new security issues such as those arising in the Arctic region.[281]

The report states: "We will seek out opportunities to work with Moscow on emerging issues, such as the future of the Arctic" and that DOD "will also enhance defense relationships and continue to work with Canada in the context of regional security, increased interaction in the Arctic, and combat operations in Afghanistan."[282]

April 2011 Change to DOD Unified Command Plan

In April 2011, President Obama assigned responsibility for the Arctic to U.S. Northern Command. Previously, U.S. Northern Command, U.S. European Command, and U.S. Pacific Command had shared responsibility for the Arctic. The April 2011 change in DOD's Unified Command Plan also assigned Alaska to U.S. Northern Command. Previously, U.S. Northern Command and U.S. Pacific Command had shared responsibility for Alaska and adjacent waters.[283]

May 2011 DOD Report to Congress

A May 2011 DOD report to Congress on Arctic operations and the Northwest Passage that was prepared at congressional direction[284] stated:

> The Arctic is warming on average twice as fast as the rest of the planet, resulting in increased human activity in the region. Although some perceive that competition for resources and boundary disputes may result in conflict in the Arctic, the opening of the Arctic also presents opportunities to work collaboratively in multilateral forums to promote a balanced approach to improving human and environmental security in the region.
>
> Strategic guidance on the Arctic is articulated in National Security Presidential Directive (NSPD) 66 / Homeland Security Presidential Directive (HSPD) 25, *Arctic Region Policy*. Additional guidance is found in *the 2010 National Security Strategy (NSS)* and the *2010 Quadrennial Defense Review (QDR)*. The overarching strategic national security objective is *a stable and secure region where U.S. national interests are safeguarded and the U.S. homeland is protected.* This objective is consistent with a regional policy that reflects the relatively low level of threat in a region bounded by nation states that have not only publicly committed to working within a common framework of international law and diplomatic engagement, but also demonstrated ability and commitment to doing so over the last fifty years.
>
> DoD will take responsible steps to anticipate and prepare for the Arctic operations of the near-(2010-2020), mid- (2020-2030), and far-term (beyond 2030). Capabilities will need to be reevaluated as conditions change, and gaps must be addressed in order to be prepared to operate in a more accessible Arctic. Key challenges include: shortfalls in ice and weather reporting and forecasting; limitations in command, control, communications, computers, intelligence, surveillance, and reconnaissance (C4ISR) due to lack of assets and harsh environmental conditions; limited inventory of ice-capable vessels; and limited

shore-based infrastructure. The key will be to address needs in step with the rate at which activity in the Arctic increases, and balance potential investments in these capabilities with other national priorities. The United States has a vital Arctic neighbor and partner in Canada, with its shared values and interests in the region. DoD will work with the Canadian Department of National Defence (DND) to ensure common Arctic interests are addressed in a complementary manner....

Existing DoD posture in the region is adequate to meet near- to mid-term U.S. defense needs. DoD does not currently anticipate a need for the construction of a deep-draft port in Alaska between now and 2020. Given the long lead times for construction of major infrastructure in the region, DoD will periodically re-evaluate this assessment as the Combatant Commanders update their regional plans on a regular basis.

The United States needs assured Arctic access to support national interests in the Arctic. This access can be provided by a variety of proven capabilities, including submarines and aircraft, but only U.S.-flagged ice-capable ships provide visible U.S. sovereign maritime presence throughout the Arctic region. This need could potentially be met by either icebreakers or ice-strengthened surface vessels, none of which are in the U.S. Navy current surface combatant inventory, but which do exist in U.S. Coast Guard's inventory in limited numbers.

Finally, significant uncertainty remains about the rate and extent of climate change in the Arctic and the pace at which human activity will increase. The challenge is to balance the risk of being late-to-need with the opportunity cost of making premature Arctic investments. Not only does early investment take resources from other pressing needs, but the capabilities would be later in their lifecycle when finally employed. Given the many competing demands on DoD's resources in the current fiscal environment, the Department believes that further evaluation of the future operating environment is required before entertaining significant investments in infrastructure or capabilities.[285]

January 2012 GAO Report Reviewing May 2011 DOD Report

A January 2012, congressionally directed[286] GAO report reviewing the May 2011 DOD report above stated the following:

While DOD has undertaken some efforts to assess the capabilities needed to meet national security objectives in the Arctic, it is unclear whether DOD will be in a position to provide needed capabilities in a timely and efficient manner because it lacks a risk-based investment strategy for addressing near-term needs and a collaborative forum with the Coast Guard for addressing long-term capability needs. DOD's [May 2011] Arctic Report acknowledges that it has some near-term gaps in key capabilities needed to communicate, navigate, and maintain awareness of activity in the region. However, DOD has not yet evaluated, selected, or implemented alternatives for prioritizing and addressing near-term Arctic capability needs. In addition, DOD and the Coast Guard have established a working group to identify potential collaborative efforts to enhance U.S. Arctic capabilities. This working group is focused on identifying potential near-term investments but not longer-term needs, and it is currently expected to be dissolved in January 2012. Uncertainty involving the rate of Arctic climate change necessitates careful planning to ensure efficient use of resources in developing Arctic needs such as basing infrastructure and icebreakers, which require long lead times to develop and are expensive to build and maintain. Without taking steps to meet near- and long-term Arctic capability needs, DOD risks making premature Arctic investments, being late in obtaining needed capabilities, or missing opportunities to minimize costs by collaborating on investments with the Coast Guard.[287]

June 2012 Senate Armed Services Committee Report Language

The Senate Armed Services Committee, in its report (S.Rept. 112-173 of June 4, 2012) on S. 3254, the FY2013 National Defense Authorization Act, stated:

> Arctic region
>
> The committee recognizes the continued importance of the Arctic region to our broader national strategy. Declining ice cover continues to open the Arctic region and a concerted, systematic, and immediate effort should be undertaken to adequately protect the United States' security, environmental, energy, economic, and natural resource interests in the Arctic.
>
> The committee commends the Department of Defense (DOD) and, in particular, U.S. Northern Command and the Department of the Navy, and other federal agencies, such as the Department of Homeland Security (DHS), through the U.S. Coast Guard, the Department of Commerce, through the National Oceanic and Atmospheric Administration and the U.S. Arctic Research Commission, for their progress thus far in studying the region and defining the capabilities required to effectively operate in and protect the domain.
>
> International and interagency collaboration is also needed to develop the necessary mapping and charting resources required for safe navigation and to promote security and economic interests. As sea ice recedes, timely weather forecasts and disaster warnings along with more baseline data will be required to conduct successful search and rescue missions. Search and rescue coordination, planning, and training for the Arctic should be thoroughly analyzed and developed to ensure forces can successfully operate in the domain.
>
> The committee recognizes the importance of DOD's involvement in interagency and international efforts to protect national security interests in the region and, accordingly, urges DOD to continue to work in concert with DHS to establish a formal chartered working group to pursue increasing Arctic capabilities in the areas of communications, maritime domain awareness, infrastructure, and presence, as was recommended by U.S. Northern Command and the U.S. Coast Guard in a joint White Paper dated March 13, 2012. The committee also urges the agencies to develop an investment strategy for funding emerging requirements in balance with a resource constrained environment. (Pages 194-195)

August 2013 Senate Appropriations Committee Report Language

The Senate Appropriations Committee, in its report (S.Rept. 113-85 of August 1, 2013) on S. 1429, the FY2014 DOD Appropriations Act, stated:

> ENGAGEMENT ON ARCTIC ISSUES
>
> The Committee notes that the physical changes in the Arctic are unprecedented in both their rate and scope of change. In addition to the economic and social concerns, numerous studies, including the 2010 Quadrennial Defense Review and the U.S. Navy Arctic Roadmap, have documented the significant impact that a rapidly changing Arctic has on National Security, to include an increasing number of vessels expected to be operating in the Arctic Ocean in the summer and fall months.
>
> The Arctic Council has grown significantly in recent years with increased influence, visibility, and membership. As of May 2013, China, Japan, Singapore, India, The Republic of Korea and Italy have been granted observer status. In 2015, the U.S. will assume its 2-year Chair of the Arctic Council, succeeding Canada in this role. In view of the increasing significance and importance of the Arctic Council, and the United States' upcoming leadership in this body, the Committee urges the Secretary of Defense to

continue to examine ways the Department can support engagement on Arctic issues, including funding for better Arctic Domain Awareness, mapping the U.S. Arctic waters, and enhanced observations and prediction of Arctic weather, ocean, and ice conditions. (Page 14)

November 2013 DOD Arctic Strategy

On November 22, 2013, DOD released a DOD strategy for the Arctic.[288] The executive summary of the document states (highlights as in the original):

The Arctic is at a strategic inflection point as its ice cap is diminishing more rapidly than projected and human activity, driven by economic opportunity—ranging from oil, gas, and mineral exploration to fishing, shipping, and tourism—is increasing in response to the growing accessibility. Arctic and non-Arctic nations are establishing their strategies and positions on the future of the Arctic in a variety of international forums. Taken together, these changes present a compelling opportunity for the Department of Defense (DoD) to work collaboratively with allies and partners to promote a balanced approach to improving human and environmental security in the region in accordance with the 2013 *National Strategy for the Arctic Region*.

Security in the Arctic encompasses a broad spectrum of activities, ranging from resource extraction and trade to activities supporting safe commercial and scientific operations to national defense. Security cooperation activities and other military-to-military forms of engagement establish, shape, and maintain international relations and the partnerships necessary to meet security challenges and reduce the potential for friction. The Department will continue to build cooperative strategic partnerships that promote innovative, affordable security solutions, and burden-sharing in the Arctic, and seek to increase opportunities with Arctic partners to enhance regional expertise and cold-weather operational experience.

The Department will continue to train and operate routinely in the region4 as it monitors the changing environment, revisiting assessments and taking appropriate action as conditions change.

This strategy identifies the Department's desired end-state for the Arctic: a secure and stable region where U.S. national interests are safeguarded, the U.S. homeland is protected, and nations work cooperatively to address challenges. It also articulates two main supporting objectives: Ensure security, support safety, and promote defense cooperation, and prepare to respond to a wide range of challenges and contingencies—operating in conjunction with other nations when possible, and independently if necessary—in order to maintain stability in the region. Finally, it identifies the ways and means the Department intends to use to achieve these objectives as it implements the National Strategy for the Arctic Region.[289]

The document also states:

U.S. national security interests in the Arctic are delineated in National Security Presidential Directive (NSPD) 66/Homeland Security Presidential Directive (HSPD) 25, Arctic Region Policy. This policy states that national security interests include such matters as missile defense and early warning; deployment of sea and air systems for strategic sealift, strategic deterrence, maritime presence, and maritime security operations; and ensuring freedom of the seas. Preserving freedom of the seas, which includes all of the rights, freedoms, and uses of the seas and adjacent airspace, including freedom of navigation and overflight, in the Arctic supports the nation's ability to exercise these rights, freedoms, and uses of the sea and airspace throughout the world, including through strategic straits.[290]

The document states that DOD

> will pursue comprehensive engagement with allies and partners to protect the homeland and support civil authorities in preparing for increased human activity in the Arctic. Strategic partnerships are the center of gravity in ensuring a peaceful opening of the Arctic and achieving the Department's desired end-state. Where possible, DoD will seek innovative, low-cost, small-footprint approaches to achieve these objectives (e.g., by participating in multilateral exercises ...). The Department [of Defense] will also evolve its infrastructure and capabilities in step with the changing physical environment in order to ensure security, support safety, promote defense cooperation, and prepare to respond to a wide range of challenges and contingencies in the Arctic in the coming decades. The Department [of Defense] will accomplish its objectives through the following ways:
>
> - Exercise sovereignty and protect the homeland;
> - Engage public and private sector partners to improve domain awareness in the Arctic;
> - Preserve freedom of the seas in the Arctic;
> - Evolve Arctic infrastructure and capabilities consistent with changing conditions;
> - Support existing agreements with allies and partners while pursuing new ones to build confidence with key regional partners;
> - Provide support to civil authorities, as directed;
> - Partner with other departments and agencies and nations to support human and environmental safety; and
> - Support the development of the Arctic Council and other international institutions that promote regional cooperation and the rule of law.[291]

The document states that challenges and risks to DOD's Arctic strategy include the possibility that projections about future access to and activity in the Arctic may prove inaccurate; the possibility that fiscal constraints may delay or deny needed investment in Arctic capabilities and curtail Arctic training; the possibility that "political rhetoric and press reporting about boundary disputes and competition for resources may inflame regional tensions"; and the possibility that "being too aggressive in taking steps to address anticipated future security risks may create the conditions of mistrust and miscommunication under which such risks could materialize."[292] Regarding the final two of these risks, the document states:

> Efforts to manage disagreements diplomatically may be hindered if the public narrative becomes one of rivalry and conflict. The Department [of Defense] will mitigate this risk by ensuring its plans, actions, and words are coordinated, and when appropriate, by engaging the press to counter unhelpful narratives with facts. The Under Secretary of Defense for Policy will monitor DoD activities, programs, and posture in the region to ensure the Department [of Defense] is sending a clear message to key audiences regarding the Department's efforts to promote security, safety, and defense cooperation....
>
> There is some risk that the perception that the Arctic is being militarized may lead to an "arms race" mentality that could lead to a breakdown of existing cooperative approaches to shared challenges. The Department [of Defense] will mitigate this risk by focusing on collaborative security approaches as outlined in the 2013 *National Strategy for the Arctic Region*, and by supporting other Federal departments and agencies where they have leadership roles. Building trust through transparency about the intent of our

military activities and participation in bilateral and multilateral exercises and other engagements that facilitate information-sharing will be a key means of addressing this risk.[293]

January 2014 Implementation Plan for National Strategy for Arctic Region

The Administration's January 2014 implementation plan for its national strategy for the Arctic region (see "Background") makes DOD the lead federal agency for one of the plan's 36 or so specific initiatives, and a supporting agency for 18 others.[294] The initiative for which DOD is designated the lead federal agency is entitled "Develop a framework of observations and modeling to support forecasting and prediction of sea ice."[295]

May 2014 House Armed Services Committee Report Language

The House Armed Services Committee, in its report (H.Rept. 113-446 of May 13, 2014) on H.R. 4435, the FY2015 National Defense Authorization Act, stated:

> Comptroller General Review of Department of Defense's Arctic Capabilities
>
> In its 2012 report on Arctic capabilities, the Government Accountability Office noted that while the Department of Defense has undertaken some efforts to assess the capabilities needed to meet national security objectives in the Arctic, it is unclear whether the Department will be in a position to provide needed capabilities in a timely and efficient manner. Pursuant to the recommendations in that GAO report that the Department of Defense develop a risk-based investment strategy and collaborate with other interagency stakeholders to identify longer term needs, the Department published an Arctic Strategy in November 2013. In that strategy, the Department identified a number of investments that will need to be made over time while noting the risk that investments in Arctic capabilities may not compete successfully against other priorities. Therefore, the committee directs the Comptroller General of the United States to provide to the congressional defense committees a review addressing:
>
> (1) The process used by the Department of Defense to identify and develop the capabilities required for the Arctic mission to include identifying the supporting force structure, personnel, training, equipment, and infrastructure;
> (2) The extent to which the Department has identified any capability gaps; and developed mitigation plans and timelines to address those gaps;
> (3) How the Department of Defense is collaborating with other agencies such as the Coast Guard to identify and address longer-term needs in the Arctic; and
> (4) Any additional information the Comptroller General deems appropriate in the context of that review.
>
> The committee directs the Comptroller General to provide a preliminary briefing to the House Armed Services Committee on the review by February 6, 2015 with a report to follow within 90 days. (Pages 123-124)

H.Rept. 113-446 also states:

> Department of Defense Arctic Collaboration with International Partners
>
> The committee notes that the Department of Defense plays a role in the U.S. National Strategy for the Arctic Region. The Department of Defense November 2013

Arctic Strategy states the Department's desired end-state for the Arctic is "a secure and stable region where U.S. national interests are safeguarded . . . and nations work cooperatively to address challenges." The strategy also states that "relationships with allies and partners are important enablers of cooperation in meeting security and defense commitments," and that these relationships "play an important role in conflict prevention, and if prevention and deterrence fail, in coordinating an international response to security and defense challenges." The committee recognizes the Department's Arctic strategy places a significant focus on collaboration with partners in the region, including the pursuit of increased multilateral collaboration with North Atlantic Treaty Organization (NATO) allies and partners, to leverage the most appropriate capabilities to meet the Department's desired end-state. The committee encourages the Department to continue fostering this collaboration with international partners, including NATO, and to keep the committee informed of its efforts. (Page 237)

June 2014 Senate Armed Services Committee Report Language

The Senate Armed Services Committee, in its report (S.Rept. 113-176 of June 2, 2014) on S. 2410, the FY2015 National Defense Authorization Act, stated:

Preparing for increased Arctic operations

The committee notes that the Arctic is an area of increasing strategic and economic importance to the United States. The opening of new maritime shipping lanes and increased access to abundant deposits of natural resources, such as oil and natural gas, will provide opportunities and challenges to U.S. national interests. The committee commends the Department of Defense (DOD) for increasing its focus on this region and encourages DOD to continue to plan for operating in what will likely become a more crowded and potentially more contested region. The committee is also interested in ensuring that DOD, in coordination with the Department of Homeland Security, focuses on developing and acquiring capabilities that will enable the United States to operate effectively in the Arctic in the coming years. The United States is an Arctic nation that stands to benefit from increased access to that region, and it is critical that our national security strategies and capabilities keep pace with our growing interests, especially as other nations increase their activity in the Arctic. (Pages 179-180)

DOD Cooperation with Canada and Other Countries

In December 2009, it was reported that "U.S. and Canadian defense officials are studying emerging gaps in their awareness of Arctic activities, seeking to boost North American Aerospace Defense [NORAD] Command's maritime-warning mission and crafting a new threat assessment for the region." The effort would reportedly involve both NORAD and a Canadian-American advisory board called the Permanent Joint Board on Defense.[296]

In May 2010, it was reported that "American and Canadian defense officials are bolstering collaboration on military exercises, investment plans and technology development related to the Arctic." The report stated that "U.S. and Canadian intelligence officials have created a classified joint "utilization" assessment for the Arctic looking out to 2020, which will be continuously updated." It also stated that "American and Canadian defense officials have just started implementing a new five-year work plan that contains specific initiatives and requires concrete deliverables, Stockton said, noting the Arctic received special attention in the plan. The plan also covers defense critical infrastructure protection, defense support to civil agencies and defense cooperation in the Americas."[297]

In December 2012, the commander of NORAD and USNORTHCOM and the commander of the Canadian Joint Operations Command signed two documents on Arctic cooperation. A December 11, 2012, statement issued by NORAD and USNORTHCOM on the signing of the documents stated in part:

> Army Gen. Charles Jacoby, Jr., commander of NORAD and U.S. Northern Command, and Lt.-Gen. Stuart Beare, Canadian Joint Operations Command commander, signed two significant documents today—the Tri Command Framework for Arctic Cooperation and the Tri-Command Training and Exercise Statement of Intent—during the 230th meeting of the Canada-U.S. Permanent Joint Board on Defense in Colorado Springs, Colo.
>
> The Tri-Command Framework for Arctic Cooperation acknowledges the Arctic is not a region of conflict and the Canadian and U.S. militaries will support other departments and agencies in response to threats and hazards in the region when requested and directed. In that context, the goal of the Framework is to promote enhanced military cooperation in the Arctic and identify specific areas of potential Tri-Command cooperation in the preparation for and conduct of safety, security and defense operations. It strengthens an already unique and mature partnership where coordination and cooperation occurs on a regular basis. The TriCommand Framework for Arctic Cooperation document is not a plan but rather outlines a process that supports the identification of opportunities for potential cooperation in the Arctic. Areas that continue to be improved, particularly in the Arctic, include planning, domain awareness, information-sharing, training and exercises, operations, capability development, and science and technology....
>
> The second document, the Tri Command Training and Exercise Statement of Intent, is aimed at enhancing joint and combined readiness in support of safety, security and defence missions through combined training and exercises and reinforcing partnerships and collaboration among the Commands. The Tri-Command Training and Exercise Directive will help to ensure a timely and coordinated response to safety, security and defence challenges to North America.
>
> The signing of the Tri Command Framework for Arctic Cooperation and the Tri-Command Training and Exercise Statement of Intent follows on a Tri Command Strategy that spells out a series of shared tasks designed to strengthen working relationships with all Tri-Command defense and security partners.[298]

In June 2013, it was reported that

> Defense chiefs representing the world's eight main Arctic nations will strengthen cooperation in marine surveillance and expand joint military exercises.
>
> Moreover, defense commanders agreed to identify and appraise the military and civilian capabilities in each country that can be used to support civilian missions in the Arctic over the next 12 months.
>
> The new strategy, following a two-day meeting of defense commanders in the coastal Greenland town of Ilulissat that ended June 12, will focus on how the eight Arctic nations can bolster defense and security cooperation in the Arctic and how military resources can be better deployed to support civilian needs across borders.[299]

In March 2014 it was reported that

> The joint Canada-U.S. North American Aerospace Defence Command wants improved surveillance systems to keep close tabs on increasing activity in the Arctic,

particularly in the region's waters, according to documents obtained by the *[Ottawa] Citizen*.

Although the installation of any new systems wouldn't take place until around 2025, the final report on what needs to be done will be presented to top military commanders on both sides of the border this spring.

The "Norad Next" initiative aims to provide direction for the alliance in the coming decades and determine what threats North America might face. It would see "the future modernization of the Norad surveillance network to provide improved multi-domain coverage, particularly in the Arctic region," according to a May 2012 briefing paper obtained by the Citizen under the Access to Information law.

The initiative is spearheaded by Norad's U.S. commander Gen. Charles Jacoby.

"A cornerstone of Gen. Jacoby's direction to his staff has been to emphasize the importance of outpacing emerging capabilities of potential adversaries," the briefing added.

Norad Next coincides with the need to replace the alliance's inventory of aging surveillance equipment. Many of its current radars reach the end of their life cycles in the 2020-2025 time frame, according to the U.S. Department of Defense.

Norad spokeswoman Capt. Jennifer Stadnyk noted that the alliance has maintained forward operating locations in the North for many years. "Additionally, the ever-increasing numbers of vessels transiting Arctic waters emphasize the need for Norad to observe, share and act on activity in that domain," she added in an email. "This will be studied during the Norad Next analysis."

Members of the Permanent Joint Board of Defence, with representatives from both countries, discussed the Norad Next concept during their meeting in December.

The initiative will make recommendations only; any decision for follow on action will have to be considered and approved by Canadian and U.S. leadership, Stadnyk added.[300]

In May 2014 it was reported that

Canadian and US military officials are looking at modernizing the North American Aerospace Defense Command's (NORAD's) surveillance capabilities as well as expanding its responsibilities to include monitoring Arctic waters.

A strategic review with various recommendations from NORAD on how to proceed is expected to be presented in the coming months to both Canada's chief of the Defence Staff, Gen. Tom Lawson, and US Army Gen. Martin Dempsey, chairman of the Joint Chiefs of Staff.

Royal Canadian Air Force Lt. Gen. Alain Parent, deputy commander of NORAD, said the review will focus NORAD toward the 2025-2030 timeframe....

But a 2012 briefing for the Canadian government noted that NORAD commander US Gen. Charles Jacoby is specifically interested in "the modernization of the NORAD surveillance network to provide improved multi-domain coverage, particularly of the Arctic region."

The five-page briefing on what is being called "NORAD Next" was forwarded to the government in June 2012 by then-Canadian Chief of the Defence Staff Gen. Walter Natynczyk.

It was obtained by Defense News and declassified under the Canadian government's access to information law.

"A cornerstone of Gen. Jacoby's direction to his staff has been to emphasize the importance of outpacing emerging capabilities of potential adversaries," the briefing noted. Those capabilities, however, have been censored from the report.

Canadian Army Capt. Jennifer Stadnyk, a spokeswoman for NORAD and US Northern Command, said the alliance's aircraft have been operating from forward locations in the Arctic for many years.

In April, NORAD aircraft conducted Operation Spring Forward, practicing a number of activities in the northern region of Canada.

But Stadnyk noted there has been an increasing amount of maritime traffic in the Arctic.

"The ever-increasing numbers of vessels transiting Arctic waters emphasize the need for NORAD to observe, share and act on activity in that domain," she said. "This will be studied during the NORAD Next analysis."

NORAD Next will make recommendations to the US and Canadian governments; they will determine which recommendations they will accept and act on, NORAD officials said.

Any recommendations that are approved will be implemented during a timeframe covering 2025 to 2030, Stadnyk said.

Stadnyk said it's not possible at this point to estimate the cost to Canada and the US of any recommendations from the NORAD review.[301]

Navy and Coast Guard in General

The Navy and Coast Guard are exploring the potential implications that increased human activities in the Arctic may have for Navy and Coast Guard required numbers of ships and aircraft, ship and aircraft characteristics, new or enlarged Arctic bases, and supporting systems, such as navigation and communication systems. The Navy and Coast Guard have sponsored or participated in studies and conferences to explore these implications, the Coast Guard has deployed cutters and aircraft into the region to perform missions and better understand the implications of operating such units there, and Navy sailors have ridden on Canadian Navy ships deploying to Arctic waters for similar reasons.

Points or themes that have emerged in studies, conferences, and deployments regarding the potential implications for the U.S. Navy and Coast Guard of diminished Arctic sea ice include but are not limited to the following:

- The diminishment of Arctic ice is creating potential new operating areas in the Arctic for Navy surface ships and Coast Guard cutters.
- U.S. national security interests in the Arctic include "such matters as missile defense and early warning; deployment of sea and air systems for strategic sealift, strategic deterrence, maritime presence, and maritime security operations; and ensuring freedom of navigation and overflight."[302]
- Search and rescue in the Arctic is a mission of increasing importance, particularly for the Coast Guard, and one that poses potentially significant operational challenges (see "Search and Rescue" above).
- More complete and detailed information on the Arctic is needed to more properly support expanded Navy and Coast Guard ship and aircraft operations in the Arctic.
- The Navy and the Coast Guard currently have limited infrastructure in place in the Arctic to support expanded ship and aircraft operations in the Arctic.
- Expanded ship and aircraft operations in the Arctic may require altering ship and aircraft designs and operating methods.
- Cooperation with other Arctic countries will be valuable in achieving defense and homeland security goals.

Navy

November 2009 Navy Arctic Roadmap

The Navy issued its first Arctic roadmap on November 10, 2009.[303] The document, dated October 2009,[304] was intended to guide the service's activities regarding the Arctic for the period FY2010- FY2014. The document has now been succeeded by the 2014-2030 Navy Arctic roadmap (see discussion below).

August 2011 Navy Arctic Environmental Assessment and Outlook Report

In August 2011, the Navy released an Arctic environment assessment and outlook report.[305] The report states:

> As the Arctic environment continues to change and human activity increases, the U.S. Navy must be prepared to operate in this region. It is important to note that even though the Arctic is opening up, it will continue to be a harsh and challenging environment for the foreseeable future due to hazardous sea ice, freezing temperatures and extreme weather. Although the Navy submarine fleet has decades of experience operating in the Arctic, the surface fleet, air assets, and U.S. Marine Corps ground troops have limited experience there. The Navy must now consider the Arctic in terms of future policy, strategy, force structure, and investments.[306]

June 2013 House Armed Services Committee Report Language

The House Armed Services Committee, in its report (H.Rept. 113-102 of June 7, 2013) on H.R. 1960, the FY2014 National Defense Authorization Act, stated:

> Navy Arctic Roadmap
>
> The committee continues to be concerned about the Department of Defense's resources and preparedness for accessing, operating in, and protecting national interests in the Arctic. The Navy currently estimates that between 2020 and 2030, the Arctic could be ice free for one month during the summer which may lead to an increase in trans-Arctic passage for vessels seeking to reduce transit distance by utilizing the Northern Sea Route and the Northwest Passage. The Navy's Strategic Objectives for the Arctic was signed in May 2010 and is referred to in both the Navy Arctic Capabilities Based Assessment, approved in September 2011, and the Navy Arctic Environmental Capabilities Based Assessment, approved in December 2012. Those objectives include ensuring Navy forces are capable and ready, contributing to the safety, stability and security in the region, safeguarding U.S. maritime interests, protecting critical infrastructure and key resources in the Arctic, and strengthening and fostering new cooperative relationships in the region. As a global Nation, the United States needs to ensure that the Navy is adequately prepared to preserve U.S. national security interests and collaborate with other Arctic nations if and when the region will be open for passage with increased traffic. The committee recognizes the importance of transparency of action without seeking to militarize the region. Therefore, the committee directs the Secretary of the Navy to provide a roadmap for future activities and costs for training and operating in the Arctic. This roadmap should be a derivative of the National Security Strategy, and shall identify proposed exercises (including table top exercises), to include the frequency, cost, and a more detailed investment strategy across the Program Objective Memorandum through fiscal years 2020 to support a timeframe leading to increased operations in the region between 2020 to 2030. Additionally, the Navy should include details regarding

international forums in which they participate. The committee further directs the Secretary of the Navy to provide this roadmap to the congressional defense committees by February 28, 2014. (Pages 130-131)

August 2013 Senate Appropriations Committee Report Language

The Senate Appropriations Committee, in its report (S.Rept. 113-85 of August 1, 2013) on S. 1429, the FY2014 DOD Appropriations Act, stated the following in a section of the report discussing the Navy's research and development account:

> *Arctic Deep Water Basing.*—The Committee commends the Department's efforts to investigate potential security and resource issues related to the opening of Arctic sea-routes. The Committee encourages the Department to continue researching a range of solutions and response, including Arctic forward basing proposals. The Committee requests the Department to provide a report to the congressional defense committees on current and future programs related to Arctic security solutions and provide an analysis of procurement and costs related to both fixed and mobile Arctic basing options. (Page 163)

January 2014 Implementation Plan for National Strategy for Arctic Region

The Administration's January 2014 implementation plan for its national strategy for the Arctic region (see "Background") mentions the Navy by name only once, as one of several agencies that will "collaborate to improve marine charting in the Arctic (Integrated Ocean and Coastal Mapping) and topographic mapping (Alaska Mapping Executive Committee)."[307] As noted above in the discussion of DOD in general, however, the January 2014 implementation plan makes DOD the lead federal agency for one of the plan's 36 or so specific initiatives and a supporting agency for 18 others.[308] The Navy will likely be a prominent participant in DOD's activities for a number of these 19 initiatives.

November 2013 DOD Arctic Strategy

The November 2013 DOD Arctic strategy (see discussion above in the section on DOD) states that "The Department of the Navy, in its role as DoD Executive Agent for Maritime Domain Awareness, will lead DoD coordination on maritime detection and tracking," and that "DoD will take steps to work with other Federal departments and agencies to improve nautical charts, enhance relevant atmospheric and oceanic models, improve accuracy of estimates of ice extent and thickness, and detect and monitor climate change indicators. In particular, the Department of the Navy will work in partnership with other Federal departments and agencies (e.g., DHS, the Department of Commerce) and international partners to improve hydrographic charting and oceanographic surveys in the Arctic."[309]

February 2014 Updated Navy Arctic Roadmap for 2014-2030

On February 24, 2014, the Navy released an updated Arctic roadmap intended to guide Navy activities regarding the Arctic for the period 2014-2030.[310] The document is the successor to the November 2009 Navy Arctic roadmap (see discussion above). The executive summary of the 2014-2030 Navy Arctic roadmap states:

> The United States Navy, as the maritime component of the Department of Defense, has global leadership responsibilities to provide ready forces for current operations and contingency response that include the Arctic Ocean. The Arctic Region remains a challenging operating environment, with a harsh climate, vast distances, and little

infrastructure. These issues, coupled with limited operational experience, are just a few substantial challenges the Navy will have to overcome in the Arctic Region. While the Region is expected to remain a low threat security environment where nations resolve differences peacefully, the Navy will be prepared to prevent conflict and ensure national interests are protected....

Navy functions in the Arctic Region are no different from those in other maritime regions; however, the Arctic Region environment makes the execution of many of these functions4 much more challenging....

In support of National and Department of Defense aims, the Navy will pursue the following strategic objectives:

- Ensure United States Arctic sovereignty and provide homeland defense;
- *Provide ready naval forces* to respond to crisis and contingencies;
- Preserve freedom of the seas; and
- *Promote partnerships* within the United States Government and with international allies and partners....

Resource constraints and competing near-term mission demands require that naval investments be informed, focused, and deliberate. Proactive planning today allows the Navy to prepare its forces for Arctic Region operations. This Roadmap emphasizes low-cost, long-lead activities that position the Navy to meet future demands. In the near to mid-term, the Navy will concentrate on improving operational capabilities, expertise, and capacity, extending reach, and will leverage interagency and international partners to achieve its strategic objectives. The Roadmap recognizes the need to guide investments by prudently balancing regional requirements with national goals....

This Roadmap provides direction to the Navy for the near-term (present-2020), mid-term (2020-2030), and far-term (beyond 2030), placing particular emphasis on near-term actions necessary to enhance Navy's ability to operate in the Arctic Region in the future. In the near-term, there will be low demand for additional naval involvement in the Region. Current Navy capabilities are sufficient to meet near-term operational needs. Navy will refine doctrine, operating procedures, and tactics, techniques, and procedures to guide future potential operations in the Arctic Region. In the mid-term, the Navy will provide support to the Combatant Commanders, United States Coast Guard, and other United States Government agencies. In the far-term, increased periods of ice-free conditions could require the Navy to expand this support on a more routine basis.[311]

Regarding "United States Navy Ways and Means for Near-Term, Mid-Term, and Far-Term Operations," the roadmap states:

Near-term: Present to 2020.

The Navy will continue to provide capability and presence primarily through undersea and air assets. Surface ship operations will be limited to open water operations in the near-term. Even in open water conditions, weather factors, including sea ice, must be considered in operational risk assessments. During shoulder seasons, the Navy may employ ice strengthened Military Sealift Command (MSC) ships to conduct Navy missions.

By 2020, the Navy will increase the number of personnel trained in Arctic operations. The Navy will grow expertise in all domains by continuing to participate in exercises, scientific missions, and personnel exchanges in Arctic-like conditions. Personnel exchanges will provide Sailors with opportunities to learn best practices from other United States' military services, interagency partners, and international allies and partners.

The Navy will refine or develop the necessary strategy, policy, plans, and requirements for the Arctic Region. Additionally, the Navy will continue to study and make informed decisions on pursuing investments to better facilitate Arctic operations. The Navy will emphasize low cost, long-lead time activities to match capability and capacity to future demands. The Navy will update operating requirements and procedures for personnel, ships, and aircraft to operate in the Region with interagency partners and allies. Through ongoing exercises, such as Ice Exercise (ICEX) and Scientific Ice Expeditions (SCICEX) research, and transits through the region by Navy submarines, aircraft and surface vessels, the Navy will continue to learn more about the evolving operating environment. The Navy will focus on areas where it provides unique capabilities and will leverage joint and coalition partners to fill identified gaps and seams.

Mid-term: 2020 to 2030.

By 2030, the Navy will have the necessary training and personnel to respond to contingencies and emergencies affecting national security. As the Arctic Ocean becomes increasingly ice-free, surface vessels will operate in the expanding open water areas. The Navy will improve its capabilities by participating in increasingly complex exercises and training with regional partners. While primary risks in the mid-term will likely be meeting search and rescue or disaster response mission demands, the Navy may also be called upon to ensure freedom of navigation in Arctic Ocean waters. The Navy will work to mitigate the gaps and seams and transition its Arctic Ocean operations from a capability to provide periodic presence to a capability to operate deliberately for sustained periods when needed.

Far-term: Beyond 2030.

In the far-term, Navy will be capable of supporting sustained operations in the Arctic Region as needed to meet national policy guidance. The Navy will provide trained and equipped personnel, along with surface, subsurface, and air capabilities, to achieve Combatant Commander's objectives. The high confidence of diminished ice coverage and navigable waterways for much of the year will enable naval forces to operate forward, ready to respond to any potential threat to national security, or to provide contingency response. Far-term risks include increased potential for search and rescue and DSCA [Defense Support of Civil Authorities], but may also require naval forces to have a greater focus on maritime security and freedom of navigation in the Region.[312]

A March 24, 2014, press report states:

The Navy has published a thorough breakdown of what it needs for future surface operations in the Arctic, from new doctrine to platform assessments and an updated cold-weather handbook for sailors.

The next step? Getting buy-in from the rest of the fleet.

With shrinking budgets and growing mission requirements elsewhere, particularly the Pacific, the service has little appetite for new tasks, experts say, especially ones that are decades away and where threats remain speculative.

"We know there is no immediate threat in the Arctic, and there are [threats] elsewhere," said Robert Freeman, a meteorologist and the spokesman for the Navy's Task Force Climate Change, which released the report last month....

The Navy study projects sea traffic in the Arctic over the coming decades, and it lays out milestones for preparing to operate in the Arctic, as well as a few general timelines. The Navy wants more sailors trained in Arctic operations by 2020. It wants to be able to respond to a national security threat in the region by 2030.

Yet as meteorologists move forward with Arctic planning, the Navy's operational side is focused on other parts of the globe.

The service is already manning more ships with fewer sailors than in past years, its top officers say. Combatant commanders across the globe meanwhile request more ships than the Navy can give them, Chief of Naval Operations Adm. Jonathan Greenert told the House Armed Services Committee earlier this month.

Arctic planning is likely to take a backseat at a time of tight budgets and the possibility of further cuts in the years ahead, Freeman and others admit. Even sending surface ships to future Arctic exercises with other countries, which the Navy did in 2010 and 2012, could be difficult as the service considers other needs, they say.

Under the study, Navy leaders aren't required to meet the timelines but must produce regular progress reports, which Freeman considers a step forward.

"It's drawing in the operational community more than they were before," he said. "It's making the operational fleet look at this. They could come back and say, 'It's too hard. We have to table this until later.' And that's fine."

The trouble with waiting is that the development of Arctic technologies, infrastructure and training take time....

The two combatant commands responsible for the region—Northern Command and European Command—have yet to outline plans for Arctic operations, and there are questions of which of the Navy's numbered fleets should cover the region....

Walter Berbrick, a professor at the Naval War College and director of its Arctic Studies Group, said that absent funding, the Navy needs to begin reaching for the "low-hanging fruit" of preparation. Sailors should "cross-deck" with foreign navies such as Norway or Denmark and report back with their experiences. Officers need to attend seminars and symposiums on the region.

"There's so much uncertainty now around the region," Berbrick said. "Our people have to get smarter."

A full capabilities assessment by the Navy—which the roadmap suggested be finished this fiscal year—would be another positive step forward, Berbrick said. When the services and Congress are ready to allocate money for the Arctic, they'll know what they need, he said.[313]

May 2014 House Armed Services Committee Report Language

The House Armed Services Committee, in its report (H.Rept. 113-446 of May 13, 2014) on H.R. 4435, the FY2015 National Defense Authorization Act, stated:

Arctic Center of Excellence

The committee notes that the Department of the Navy released its Arctic Roadmap for 2014– 2030, which provides near-term, midterm, and far-term direction to enhance the Navy's ability to operate in the Arctic region. The committee notes the roadmap includes an implementation plan that provides a timeline and identifies organizations to lead specific actions recommended by the roadmap. One of these actions includes identifying requirements in fiscal year 2015 to establish an Arctic Center of Excellence, as well as developing an Arctic engagement plan focusing on partnerships with international, interagency and private sector stakeholders that enhance security. The committee believes that the establishment of such a center would support the Navy's focus areas identified in the roadmap. In developing the requirements for such a center, the committee encourages the Department to consider, among other things, how such a center could support Arctic-related training, operations, maritime domain awareness, scientific research, and technology development. When considering potential locations the Department should consider candidates that can effectively bring together elements from the Department of Defense, Department of Homeland Security, academia, and other public and private stakeholders who can contribute to advancing U.S. interests in the Arctic. (Pages 121-122)

Additional Navy Statements

In a February 2009 journal article, the Oceanographer of the Navy stated:

> Preserving freedom of navigation in the region is an important tenet of U.S. policy. The NSR [Northern Sea Route], however, is a contested waterway, with Russian claims of sovereignty competing against U.S. and European Union insistence that it is an international strait available to all nations, subject to mutually recognized terms. Another potential [Arctic] transoceanic shipping route may be the Northwest Passage, which extends from the Atlantic through Baffin Bay and the Canadian Archipelago and into the Pacific by way of the Bering Strait. Canada claims sovereignty over the waters of the Canadian Archipelago, although the United States and the European Union claim that the Northwest Passage also constitutes an international strait which allows right of innocent passage....
>
> Aside from access and right of passage, the Navy and Coast Guard, in particular, must also be concerned with strategic choke points such as the Bering Strait, Canada's Queen Elizabeth Islands in the Northwest Passage, and Russia's Severnaya Zemlya and New Siberian Islands in the Northern Sea Route. These narrow passages offer some protection from persistent ice blockage, but they are also vulnerable to control or blockade by adversaries that would significantly disrupt potential commercial shipping and oil transport....
>
> U.S. naval interests will face new challenges in an increasingly ice-free Arctic with a strategic objective to understand potential threats to the United States from the maritime domain. As throughout the global commons, the U.S. Navy must be aware of activities that could be harmful to national security interests in a region that will, no doubt, see fewer barriers to access by potential adversaries in the future. National and homeland security interests pertinent to the U.S. Navy in the region would include early warning/missile defense; maritime presence and security; and freedom of navigation and over-flight....
>
> The region is primarily a maritime domain and the U.S. Navy of the future must be prepared to protect sea lines of communication supporting maritime commerce and other national interests—including national security—there. In addition to thinking through how we adjust our shipbuilding emphasis to support such operations, the Navy should also be thinking strategically about building the necessary infrastructure to provide logistic support for Arctic patrols, search and rescue capabilities, and shore-based support activities.
>
> To ensure complete maritime domain awareness in the region, and to provide our forces a competitive advantage, it will be necessary to have comprehensive knowledge of the physical environment. Data must be obtained by a suite of remote sensors (satellites, radars), autonomous sensors (data buoys, unmanned vehicles), and manned sensors (shipboard, coastal observing stations). Computer-based ocean and atmospheric models must be adjusted to the geophysical peculiarities of high latitudes. Communication lines for data exchange and reach-back processing at high-performance computing production centers must be robust and reliable. To ensure safety of navigation, we will also need to conduct more high-resolution bottom surveys and increase the scrutiny we place on sea ice conditions.
>
> The Navy relies on its international and interagency partners for assistance to ensure success of maritime domain awareness and maritime security missions. To meet the demands of national security in the changing northern environment, strengthening mechanisms for cooperation among the regional nations and U.S. agencies must remain a

high priority. Like everywhere else in the world, sound national security in the Arctic will require strategic access, military mobility, safe navigation, unimpeded maritime transportation, improved homeland security, and responsible, sustainable use of ocean and coastal resources. International and interagency agreements and partnerships are vital to incorporating these essential elements into a viable national security policy and will be critical for resolving future naval challenges of a changing Arctic.[314]

In a spring 2010 journal article, the Oceanographer of the Navy (the successor to the officer who authored the above journal article) and a coauthor stated that "Navy surface vessels are able to operate up to the marginal ice zone but will require ice-strengthening to operate in higher ice conditions; Navy aircraft are capable of operating in the Arctic, but the lack of divert fields limits their duration and range."[315] The article also stated:

> While the Navy has a rich history in the Arctic, several challenges must be met to ensure successful operations in the future. These include the lack of support infrastructure and logistics support, environmental hazards such as drifting sea ice and icing on exposed surfaces, and communications difficulties. Antiquated nautical charts, drifting ice, low visibility, and the paucity of electronic and visual navigation aids hinder safety of navigation. A lack of coastal installations also contributes to the difficulty of search and rescue (SAR) operations. The only American-owned deepwater port near the Arctic basin is Dutch Harbor, in the Aleutian Islands.[316]

On June 16, 2011, in remarks at a seminar on the Arctic, Admiral Gary Roughead, the Chief of Naval Operations, stated:

> Well in my mind, there is a phenomenal event taking place on the planet today, and that is what I call the opening of the Fifth Ocean; that's the Arctic Ocean. We haven't had an ocean open on this planet since the end of the Ice Age. So if this is not a significant change that requires new, and I would submit, brave thinking on the topic, I don't know what other sort of physical event could produce that.
>
> We in the Navy a couple of years ago began to look at how we as a Navy serve the nation.... So as we began to look at the planet, we realized that there were significant changes taking place; opening of the Fifth Ocean; physical changes; changes in population as they compressed into the littoral areas.
>
> And so we established, I established something that I call Task Force Climate Change. Primarily focused on the Arctic because of the massive changes that are taking place up there, but it also takes into account what is happening on the rest of the planet and where we as a Navy think we have to be in the future. And then from that Task Force Climate Change, we developed what we call the Arctic road map....
>
> The way we see things taking place is the first of moves, which will largely be in the area of resources. Fishing stocks will move with the water temperature, they will start to drift farther north, that will take fishing fleets farther north and there is a set of responsibilities that we have with other nations on making sure that that activity is taking place lawfully.
>
> Then we believe the next step will be the extraction of resources. So what do we have to be prepared to do as a Navy, as a military particularly aligned closely with the Coast Guard on this.
>
> And then in about 20 years time, 25 years time, the Arctic becomes a profitable sea route from Asia to Europe over the top of the planet. If you look at some of the estimates from shippers, some may actually be here. That is about a million dollars a trip someone saves. That is not insignificant. And so it will become a busy place as transportation starts

to take place. That is how we envision the changes that are taking place, how we are looking at them, and the road map that we put in place is based on the best science as we know today. And we are committed to providing the resources the nation needs from the naval perspective to meet our needs.[317]

Four days later, in remarks to a symposium on the impacts on naval and maritime operations of an ice-diminishing Arctic, Admiral Roughead stated:

Since this forum began in 2001, our Navy has benefitted from the development and the discussion of the Arctic Symposium, as it has improved our appreciation greatly for an ice-diminished Arctic and the security implications that will follow from that physical change that will take place.

This broader examination is something that we've endeavored to focus on systematically since we stood up our Task Force Climate Change in May 2009, and issued our Arctic Roadmap that same year. And rather than all the answers, what our efforts have given us is an appreciation of just how dynamic the study of the region is and surely will be in the future....

... at a fundamental level, the trends point undeniably towards a new venue of operations and responsibility for our global Navy—for preserving American interests in free and fair access there—and in light of this we remain committed to preparing exhaustively for the challenges and especially for the opportunities that are going to exist in an ice-diminishing Arctic.

The U.S. Navy's interests in the Arctic are not new, of course. We have many decades of experience with exploration and, indeed, episodic operations in the waters of the Arctic Circle....

But never has our interest encountered the confluence of trends, as projected by the U.S. Geological Survey in 2008 and the National Research Council this past March, that promises to change the Arctic so pervasively, and in so doing affect the global environment for which we plan and program our future fleet....

In projecting the impact of climate change in the northern latitudes, however, I'm reminded of what [then-NOAA Administrator] Dr. [Jane] Lubchenco observed just this past March, when she said, "what happens in the Arctic does not stay there." The trends we discuss here, in a similar timeframe, promise more disruption and disorder in a world whose population is growing rapidly, and moving to megacities on or near the coasts of almost every continent. The prospects of sea level rise, for some megacities, or the coral islands of the Maldives, are similarly daunting. We also have to consider the likely frictions that arise as fishing stocks migrate with changing sea temperatures, and the very real possibility that conflicts in the future will be fought over access to dwindling natural supplies of fresh water.

It is because of these projections that our Navy is preparing for increased demand, both in the region—where we will maintain our access and uphold the freedom of navigation as a global good—and beyond, where we expect developments to expose the costs of our national reluctance on the Law of the Sea convention and to test our present understanding of customary legal guarantees to the very freedoms behind our global operations today. We are considering the technical requirements for polar operations to support our strategic objective of a safe, stable, and secure Arctic region where our national interests are safeguarded— namely, how and when to build forces capable and competent for the harsh northern climes.

We also remain well aware of how important maritime partnerships will be in addressing capacity concerns as we seek to reinforce universal values in the global commons. Some speak of a changing Arctic in terms of either a race or a zero-sum game. As far as the U.S. Navy is concerned, it is neither. Rather, we see it as an opportunity to extend to the 'Fifth Ocean' the principles that have benefitted all peaceful nations in the other four. Secretary of State Clinton's attendance at the meeting of Arctic Council ministers last month in Nuuk, Greenland, signals the United States" level of interest in the region, as well as our belief in the power of international cooperation to advance mutual interests in the free use of the maritime domain. It was also inherently practical in that it forged the much-needed agreement among Arctic states to improve our coordination on search and rescue.

As we stated in our Arctic Roadmap, the Navy will continue to pursue cooperative relationships around such areas of common interest, whether mission enablers like Maritime Domain Awareness, core naval capabilities such as humanitarian assistance and disaster response, or intra-governmental relationships like defense support to civil authorities. Exercises such as the Canadian Navy's 'Operation Nanook,' which we observed in 2009 and had the privilege of participating in last August alongside our Coast Guard counterparts, enhance our preparedness in several of these areas, and lay the basis from which we might extend the same maritime partnership and interoperability we have come to value so highly in our global operations. Joint exercises such as the U.S. Pacific Command's 'Northern Edge'—which started just outside of the Arctic Circle one week ago today and will continue throughout this week – ensure that we as a Navy remain ready to partner at the high end of operations as required.[318]

An April 28, 2012, press report states:

As global warming opens the Arctic Ocean to commercial and industrial traffic, the U.S. Navy is pushing to catch up with Russia, Canada and even Denmark in its Arctic ability. If a crisis were to happen now, the Navy lacks the ability to act in the Arctic without the help of one of those countries or the Coast Guard.

Last year, the Navy asked the War Gaming Department of the U.S. Naval War College to find out what the Navy needs for sustained operations in the Arctic.

In the resulting 2011 Fleet Arctic Operations Game, the Navy learned how big its Arctic shortcomings are. As a force, the Navy lacks everything from bases and Arctic-capable ships to reliable communications and cold-weather clothing....

The game's conclusions: the Navy is not adequately prepared to conduct long-term maritime Arctic operations; Arctic weather conditions increase the risk of failure; and most critically, to operate in the Arctic, the Navy will need to lean on the U.S. Coast Guard, countries like Russia or Canada, or tribal and industrial partners.

To sustain operations in the Arctic, the Navy needs ice-capable equipment, accurate and timely environmental data, personnel trained to operate in extreme weather, and better communications systems. Much of the environmental data will come from other Arctic nations....

Navy officials understand the need to conduct exercises in the Arctic so they can get ready for the real thing, but they don't have a strategy.

"We are the only Arctic nation without an Arctic strategy," said U.S. Navy Cmdr. Blake McBride, Arctic Affairs officer for Task Force Climate Change. "The Coast Guard and Department of Defense are working on a strategy to help answer the issue, and advocate for capabilities."

Aside from signing National Security Presidential Directive 66, which requires the U.S. to have a presence in the Arctic, the Arctic hasn't been a priority for the U.S. government, largely because there isn't an immediate military threat.

"It's becoming a higher priority, but we don't make our own priorities," McBride said. "We don't foresee a military threat in the Arctic, but it doesn't mean we will not need to be able to operate there."

The Navy's future plans to conduct operations in the Arctic largely depend on the budget.

"It's all about the money," McBride said. "If you don't have the budget or funds to invest in manpower and equipment then you don't have anything."[319]

Coast Guard

Coast Guard High Latitude Study Provided to Congress in July 2011

In July 2011, the Coast Guard provided to Congress a study on the Coast Guard's missions and capabilities for operations in high-latitude (i.e., polar) areas. The study, commonly known as the High Latitude Study, is dated July 2010 on its cover. The High Latitude Study concluded the following:

> [The study] concludes that future [Coast Guard] capability and capacity gaps will significantly impact four [Coast Guard] mission areas in the Arctic: Defense Readiness, Ice Operations, Marine Environmental Protection, and Ports, Waterways, and Coastal Security. These mission areas address the protection of important national interests in a geographic area where other nations are actively pursuing their own national goals. U.S. national policy and laws define the requirements to assert the nation's jurisdiction over its territory and interests; to ensure the security of its people and critical infrastructure; to participate fully in the collection of scientific knowledge; to support commercial enterprises with public utility; and to ensure that the Arctic environment is not degraded by increased human activity.
>
> The Coast Guard's ability to support Defense Readiness mission requirements in the Arctic is closely linked to DoD responsibilities. The Coast Guard presently possesses the only surface vessels capable of operating in ice-covered and ice-diminished waters. The Coast Guard supports (1) DoD missions such as the resupply of Thule Air Base in Greenland and logistics support (backup) for McMurdo Station in Antarctica and (2) Department of State (DoS) directed Freedom of Navigation Operations. These unique Coast Guard capabilities have been noted by the Joint Chiefs of Staff, the Navy's Task Force Climate Change, and the recently issued Naval Operations Concept 2010.
>
> The common and dominant contributor to these significant mission impacts is the gap in polar icebreaking capability....[320]
>
> Other capability gaps contributing to the impact on Coast Guard ability to carry out its missions in the Arctic include:
>
> - Communications System Capability – Continuous coverage along Alaska's West Coast, the Bering Strait, and throughout the North Slope is required for exchanging voice and data communications with Coast Guard units and other government and commercial platforms offshore.
> - Forward Operating Locations - No suitable facilities currently exist on the North Slope or near the Bering Strait with facilities sufficient to support extended aircraft servicing and maintenance. Aircraft must travel long distances and expend significant time transiting to and from adequate facilities. This gap reduces on-scene presence and capability to support sustained operations in the region.
> - Environmental response in ice-covered waters - The technology and procedures for assessment and mitigation measures for oil spills in ice-covered waters are not fully developed or tested.

Capability gaps in the Arctic region have moderate impacts on [the Coast Guard's] Aids to Navigation (AtoN), Search and Rescue (SAR), and Other Law Enforcement (OLE) missions. Both AtoN and SAR involve the safety of mariners and will gain more importance not only as commerce and tourism cause an increase in maritime traffic, but as U.S. citizens in northern Alaska face more unpredictable conditions. Performance of OLE will be increasingly necessary to ensure the integrity of U.S. living marine resources from outside pressures....

In addition to the assessment of polar icebreaking needs, the Arctic mission analysis examined a set of theoretical mixes (force packages) of Coast Guard assets consisting of icebreakers, their embarked helicopters, and deployment alternatives using aviation forward operating locations in Arctic Alaska....

All [six] of the force mixes [considered in the study] add assets to the existing Coast Guard Alaska Patrol consisting of (1) a high-endurance cutter (not an icebreaker) deployed in the Bering Sea carrying a short range recovery helicopter, and (2) medium range recovery helicopters located at Kodiak in the Gulf of Alaska, and seasonally deployed to locations in Cold Bay and St. Paul Island....

These force packages and associated risk assessment provide a framework for acquisition planning as the Coast Guard implements a strategy for closing the capability gaps. By first recapitalizing the aging icebreakers, the Coast Guard provides a foundation for buildout of these force mixes. In addition to the cost of the icebreakers, the force packages require investment in forward operating locations and in medium range helicopters. The mission analysis reports developed rough order-of-magnitude cost estimates for forward operating locations at approximately $36M [million] each and for helicopters at $9M each....

The analysis shows that the current Coast Guard deployment posture is not capable of effective response in northern Alaska and that response may be improved through a mix of deployed cutters, aircraft, and supporting infrastructure including forward operating locations and communications/navigation systems.[321]

May 2013 Coast Guard Arctic Strategy

On May 21, 2013, the Coast Guard released a strategy document for the Arctic.[322] The executive summary of the document states in part:

The U.S. Coast Guard, as the maritime component of the U.S. Department of Homeland Security (DHS), has specific statutory responsibilities in U.S. Arctic waters. This strategy outlines the ends, ways, and means for achieving strategic objectives in the Arctic over the next 10 years. The Coast Guard is responsible for ensuring safe, secure, and environmentally responsible maritime activity in U.S. Arctic waters. Our efforts must be accomplished in close coordination with DHS components, and involve facilitating commerce, managing borders, and improving resilience to disasters.

The Coast Guard's current suite of cutters, boats, aircraft, and shore infrastructure must meet a number of near-term mission demands. The Coast Guard employs mobile command and control platforms such as large cutters and ocean-going ice-strengthened buoy tenders, as well as seasonal air and communications capabilities through leased or deployable assets and facilities. These mobile and seasonal assets and facilities have proven to be important enablers for front-line priorities in the region, including search and rescue operations, securing the maritime border, collecting critical intelligence, responding to potential disasters, and protecting the marine environment....

Although winter sea travel is still severely limited due to extensive ice coverage across the region, recent summer and early autumn sea ice extent record lows have made seasonal maritime navigation more feasible. Economic development, in the forms of resource extraction, adventure tourism, and trans-Arctic shipping drives much of the current maritime activity in the region.

[Oil and gas exploration] activities [in the region] bring risk, which can be mitigated through appropriate maritime governance. Additionally, tourism is increasing rapidly in the Arctic. Due to undeveloped shore-based infrastructure, much of the increased tourism is expected to involve transportation via passenger vessel, further increasing near- and offshore activities in Arctic waters.

This document outlines three strategic objectives in the Arctic for the U.S. Coast Guard over the next 10 years:

- Improving Awareness
- Modernizing Governance
- Broadening Partnerships

Improving Awareness: Coast Guard operations require precise and ongoing awareness of activities in the maritime domain. Maritime awareness in the Arctic is currently restricted due to limited surveillance, monitoring, and information system capabilities. Persistent awareness enables identification of threats, information-sharing with front-line partners, and improved risk management. Improving awareness requires close collaboration within DHS, as well as with the Departments of State, Defense, Interior, the National Science Foundation and other stakeholders to enhance integration, innovation, and fielding of emerging technologies. The Intelligence Community and non-federal partners are also vital stakeholders.

Modernizing Governance: The concept of governance involves institutions, structures of authority, and capabilities necessary to oversee maritime activities while safeguarding national interests. Limited awareness and oversight challenge maritime sovereignty, including the protection of natural resources and control of maritime borders. The Coast Guard will work within its authorities to foster collective efforts, both domestically and internationally, to improve Arctic governance. In so doing, the Coast Guard will review its own institutions and regimes of governance to prepare for future missions throughout the Arctic.

Broadening Partnerships: Success in the Arctic requires a collective effort across both the public and private sectors. Such a collective effort must be inclusive of domestic regulatory regimes; international collaborative forums such as the Arctic Council, International Maritime Organization (IMO), and Inuit Circumpolar Council; domestic and international partnerships; and local engagements in Arctic communities focusing on training and volunteer service. Success in the Arctic also depends upon close intergovernmental cooperation to support national interests, including working closely within DHS, as well as with the Department of State, Department of Interior and other Federal partners as the U.S. prepares to assume Chairmanship of the Arctic Council in 2015.

Beyond these three strategic objectives, there are a number of additional factors that will position the Coast Guard for long-term success. These factors include building national awareness of the Arctic and its opportunities, strengthening maritime regimes, improving public-private relationships through a national concept of operations, seeking necessary authorities, and identifying future requirements and resources to shape trends favorably. This strategy outlines a number of priorities, ranging from capabilities and requirements to advances in science and technology that will facilitate our Nation's success in the region. Specifically, the strategy advocates to leverage the entire DHS enterprise and component capabilities to secure our borders, prevent terrorism, adapt to changing environmental conditions, enable community resilience and inform future policy.

Operating in the Arctic is not a new venture for the Coast Guard. However, adapting to changing conditions will require foresight, focus, and clear priorities. This strategy will ensure we attain the aim of safe, secure, and environmentally responsible maritime activity in the Arctic by improving awareness, modernizing governance, and broadening partnerships to ensure long-term success.[323]

A June 2013 press report stated:

Arctic sea lanes may be staying open longer and traffic in the frigid waters may be increasing, but the Coast Guard's new 10-year plan for Arctic missions won't send more Coasties to the region or establish a year-round presence, the service's top officer said.

The service's seasonal, mobile approach has been "tested and proven ... over the last several years," Commandant Adm. Bob Papp said when he released the Coast Guard's Arctic Strategy on May 21. "That will be our approach in the coming decade."

The service will not increase deployment time to the Arctic, continuing to rely on standard six-month cutter deployments to maintain a presence, Papp said.

The national security cutters carry supplies, small boats and helicopters, and have the communication and intelligence capabilities needed for a sustained offshore presence in the ice-free waters, he said. They patrol the Arctic for the summer season, typically from June to September, depending on how long the waters are ice-free, said Lt. Veronica Colbath, a spokeswoman for District 17, which covers Alaska.

Using the cutter fleet instead of establishing bases over the next 10 years will save money and allow for greater flexibility, Papp said. The strategy also stresses expanding partnerships with domestic and international regulatory bodies monitoring the Arctic, including the Arctic Council, International Maritime Organization and the Inuit Circumpolar Council.

With long-term forecasts showing the region will become ice-free for longer periods, Papp said that at some point the Coast Guard might need a full-time sector, based in Barrow, Alaska.

However, that's not part of the current strategy.

"We are not going to permanently station anyone up there," Papp said.[324]

July 2013 Senate Appropriations Committee Report Language

The Senate Appropriations Committee, in its report (S.Rept. 113-77 of July 18, 2013) on H.R. 2217, the FY2014 Department of Homeland Security Appropriations Act, stated:

INFRASTRUCTURE AND RESPONSE CAPABILITIES IN THE ARCTIC

The Committee is concerned about the lack of assets available for the Coast Guard's Arctic mission. No later than 120 days after the date of enactment of this act, the Commandant is directed to submit a report to the Committee comparing the costs of facility renovations to homeport and support an NSC [a National Security Cutter] in Alaska with the annual costs of transit time to Alaska area of operations for deployments and days lost to casualty repairs.

ARCTIC STRATEGY

The Coast Guard recently released its Arctic Strategy, which looks at the changing conditions in the region and contemplates future requirements. The strategy outlines three broad objectives: improving awareness; modernizing governance; and broadening partnerships. To carry out these objectives over the long term, the Coast Guard next needs to develop an implementation plan, including the identification of necessary capabilities, requirements, authorities, and resources. Therefore, the Coast Guard is directed to submit an Arctic strategy implementation plan no later than 120 days after the date of enactment of this act.

INTERNATIONAL MARITIME ORGANIZATION

The Committee wants to ensure that as activities in the Arctic expand, necessary response capabilities exists in the region. No later than 90 days after the date of enactment of this act, the Secretary is directed to submit a report to the Committee on activities that have occurred pursuant to section 307(b) of Public Law 111–281 and how the Department is meeting the requirements being developed in the agreements envisioned by that subsection.[325] (Pages 80- 81)

January 2014 Implementation Plan for National Strategy for Arctic Region

The Administration's January 2014 implementation plan for its national strategy for the Arctic region (see "Background") makes "Department of Homeland Security (United States Coast Guard)" the lead federal agency for six of the plan's 36 or so specific initiatives, and a supporting agency for 13 others.[326] The six initiatives where the Coast Guard is designated the lead federal agency include:

- enhance Arctic domain awareness;
- improve hazardous material spill prevention, containment, and response;
- promote Arctic oil pollution preparedness, prevention, and response internationally;
- enhance Arctic search and rescue;
- expedite International Maritime Organization (IMO) Polar Code development and adoption; and
- promote Arctic waterways management.[327]

For the second initiative above—"Improve Hazardous Material Spill Prevention, Containment, and Response"—the Coast Guard shares lead-agency status with the Environmental Protection Agency (EPA), with the Coast Guard being the lead federal agency for open ocean and coastal spills, and EPA being the lead federal agency for inland spills.[328]

January 2009 Arctic Policy Directive (NSPD 66/HSPD 25)

As noted earlier (see "January 2009 Arctic Policy Directive (NSPD 66/HSPD 25)" in "Background"), the Obama Administration is currently operating under the January 2009 Arctic region policy directive (NSPD 66/HSPD 25) issued by the George W. Bush Administration.[329] Potential oversight questions include but not are limited to the following:

- Is NSPD 66/HSPD 25 a suitable statement of U.S. policy for the Arctic region?[330]
- Although the Obama Administration is currently operating under NSPD 66/HSPD 25, does the Obama Administration fully agree with all parts of it? If not, with which parts does it not fully agree?
- Does the Obama Administration intend to eventually conduct a review of NSDP 66/HSPD 25? If so, what is the Administration's schedule for conducting and releasing the results of that review?

In connection with the above questions, a March 2013 report from the Center for Strategic and International Studies (CSIS) on U.S. interests and U.S. government actors in the Arctic stated:

> NSPD-66/HSPD-25 contains a rigorous implementation schedule for the federal government. Yet, as this directive passes its fourth anniversary, many implementation strategies have not been put into effect. That said, several of the policy priorities and implementation strategies are redundant. For example, to meet U.S. security interests in the Arctic, an interagency group under the leadership of the departments of State, Defense, and Homeland Security is requested to "develop greater capabilities and capacity [...] to protect U.S. air, land and sea borders" and "increase Arctic maritime domain awareness." Under maritime transportation, these same agencies, with the departments of Transportation and Commerce, are to "determine basing and logistics support requirements, including necessary airlift and icebreaking capabilities; and improve plans and cooperative agreements for search and rescue." These overlapping mandates raise the question whether this strategy is a security function, an act of commerce—or both—and who ultimately is accountable for implementing the strategy.
>
> Such policy overlap and redundancy accurately reflects U.S. Arctic policymaking today and unfortunately reflects poorly on the interagency process following the release of NSPD-66. A successful future policy demands a more streamlined and prioritized process.
>
> It is time to update NSPD-66. Other Arctic coastal states emerged with Arctic policy statements around the same time as the United States did, in the 2007 to 2009 timeframe. Since issuing these statements, however, most other Arctic nations have updated and further refined their whole-of-government strategies. Unfortunately, the United States has failed to do so. Separate federal agencies and departments have developed, or will be developing, their own separate strategies based on NSPD-66, but this effort has been uneven at best.... The Department of Defense report on Arctic Operations and the Northwest Passage was congressionally mandated by the fiscal year 2011 National Defense Authorization Act. However, there is no such mandate for other U.S. government agencies and, consequently, most other agencies have yet to develop their own implementation strategy stemming from NSPD-66.
>
> What would an updated U.S. Arctic strategy require? The seven policy areas identified in NSPD-66 remain relevant, but these areas must be defined more clearly.
>
> First and foremost, the United States must create a long-term economic strategy for the American Arctic. The first component of an Arctic economic strategy must be an energy, mineral resource, and infrastructure strategy. NSPD-66 states that "Energy development in the Arctic region will play an important role in meeting growing global energy demand." How large a role? At present, U.S. energy strategy consists of a five-year offshore licensing and permitting plan that ends in 2017. The government needs to define its long-term offshore and onshore energy strategy for Alaska, answering such questions as whether America's Arctic energy resources are intended to meet U.S. demand or to be exported to Asian markets. Other questions in need of answers include the status of port, pipeline, and liquid natural gas infrastructure; whether methane hydrates are viable; and whether energy development can be pursued in an environmentally sustainable way in such a fragile environment.
>
> Based on the answers to these questions, America's Arctic economic strategy must also include a detailed maritime transportation and infrastructure strategy. This strategy would ideally be built around ecosystem-based management. It is clear that a future U.S. Arctic maritime or infrastructure strategy will likely be a public-private partnership, as the private sector will provide significantly more financial and physical resources than will the U.S. government.[331]

May 2013 Arctic Strategy and January 2014 Implementation Plan

As noted earlier (see "Background"), the Administration on May 10, 2013, released a national strategy for the Arctic, and on January 30, 2014, released an implementation plan for that strategy. Potential oversight questions include but not are limited to the following:

- Does the May 2013 Arctic strategy document correctly identify U.S. interests in the Arctic and principal U.S. lines of effort for the region?
- Does the May 2013 Arctic strategy document adequately balance goals for the region against potential resources available for pursuing those goals, and provide an adequate guide for prioritizing the goals in a situation of constrained resources?
- Does the January 2014 implementation plan correctly identify the principal initiatives needed to implement the Arctic strategy document?
- Does the January 2014 implementation plan correctly identify the next steps to be taken for implementing the various initiatives? Does it provide adequate metrics for measuring progress in achieving the initiatives? Does it select the correct federal agencies to act as lead agencies and supporting agencies for achieving them?
- Does the January 2014 implementation plan adequately identify risks and challenges in implementing the various initiatives, including those related to potential limits on resources? Does it provide an adequate guide for prioritizing the initiatives in a situation of constrained resources?
- Who in the executive branch has been designated as the person responsible for ensuring that the various goals in the January 2014 implementation plan are implemented in a timely manner?[332]

APPENDIX A. ARCTIC RESEARCH AND POLICY ACT (ARPA) OF 1984 (TITLE I OF P.L. 98-373)

The text of the Arctic Research and Policy Act (ARPA) of 1984 (Title I of P.L. 98-373 of July 31, 1984)[333] is as follows:

TITLE I – ARCTIC RESEARCH AND POLICY
SHORT TITLE
SEC. 101. This title may be cited as the "Arctic Research and Policy Act of 1984".

FINDINGS AND PURPOSES
SEC. 102. (a) The Congress finds and declares that-

(1) the Arctic, onshore and offshore, contains vital energy resources that can reduce the Nation's dependence on foreign oil and improve the national balance of payments;

(2) as the Nation's only common border with the Soviet Union, the Arctic is critical to national defense;

(3) the renewable resources of the Arctic, specifically fish and other seafood, represent one of the Nation's greatest commercial assets;

(4) Arctic conditions directly affect global weather patterns and must be understood in order to promote better agricultural management throughout the United States;

(5) industrial pollution not originating in the Arctic region collects in the polar air mass, has the potential to disrupt global weather patterns, and must be controlled through international cooperation and consultation;

(6) the Arctic is a natural laboratory for research into human health and adaptation, physical and psychological, to climates of extreme cold and isolation and may provide information crucial for future defense needs;

(7) atmospheric conditions peculiar to the Arctic make the Arctic a unique testing ground for research into high latitude communications, which is likely to be crucial for future defense needs;

(8) Arctic marine technology is critical to cost-effective recovery and transportation of energy resources and to the national defense;

(9) the United States has important security, economic, and environmental interests in developing and maintaining a fleet of icebreaking vessels capable of operating effectively in the heavy ice regions of the Arctic;

(10) most Arctic-rim countries, particularly the Soviet Union, possess Arctic technologies far more advanced than those currently available in the United States;

(11) Federal Arctic research is fragmented and uncoordinated at the present time, leading to the neglect of certain areas of research and to unnecessary duplication of effort in other areas of research;

(12) improved logistical coordination and support for Arctic research and better dissemination of research data and information is necessary to increase the efficiency and utility of national Arctic research efforts;

(13) a comprehensive national policy and program plan to organize and fund currently neglected scientific research with respect to the Arctic is necessary to fulfill national objectives in Arctic research;

(14) the Federal Government, in cooperation with State and local governments, should focus its efforts on the collection and characterization of basic data related to biological, materials, geophysical, social, and behavioral phenomena in the Arctic;

(15) research into the long-range health, environmental, and social effects of development in the Arctic is necessary to mitigate the adverse consequences of that development to the land and its residents;

(16) Arctic research expands knowledge of the Arctic, which can enhance the lives of Arctic residents, increase opportunities for international cooperation among Arctic-rim countries, and facilitate the formulation of national policy for the Arctic; and

(17) the Alaskan Arctic provides an essential habitat for marine mammals, migratory waterfowl, and other forms of wildlife which are important to the Nation and which are essential to Arctic residents.

(b) The purposes of this title are-

(1) to establish national policy, priorities, and goals and to provide a Federal program plan for basic and applied scientific research with respect to the Arctic, including natural resources and materials, physical, biological and health sciences, and social and behavioral sciences;

(2) to establish an Arctic Research Commission to promote Arctic research and to recommend Arctic research policy;

(3) to designate the National Science Foundation as the lead agency responsible for implementing Arctic research policy; and

(4) to establish an Interagency Arctic Research Policy Committee to develop a national Arctic research policy and a five year plan to implement that policy.

ARCTIC RESEARCH COMMISSION

SEC. 103. (a) The President shall establish an Arctic Research Commission (hereafter referred to as the "Commission").

(b)(1) The Commission shall be composed of five members appointed by the President, with the Director of the National Science Foundation serving as a nonvoting, ex officio member. The members appointed by the President shall include-

(A) three members appointed from among individuals from academic or other research institutions with expertise in areas of research relating to the Arctic, including the physical, biological, health, environmental, social, and behavioral sciences;

(B) one member appointed from among indigenous residents of the Arctic who are representative of the needs and interests of Arctic residents and who live in areas directly affected by Arctic resource development; and

(C) one member appointed from among individuals familiar with the Arctic and representative of the needs and interests of private industry undertaking resource development in the Arctic.

(2) The President shall designate one of the appointed members of the Commission to be chairperson of the Commission.

(c)(1) Except as provided in paragraph (2) of this subsection, the term of office of each member of the Commission appointed under subsection (b)(1) shall be four years.

(2) Of the members of the Commission originally appointed under subsection (b)(1)-

(A) one shall be appointed for a term of two years;

(B) two shall be appointed for a term of three years; and

(C) two shall be appointed for a term of four years.

(3) Any vacancy occurring in the membership of the Commission shall be filled, after notice of the vacancy is published in the Federal Register, in the manner provided by the preceding provisions of this section, for the remainder of the unexpired term.

(4) A member may serve after the expiration of the member's term of office until the President appoints a successor.

(5) A member may serve consecutive terms beyond the member's original appointment.

(d)(1) Members of the Commission may be allowed travel expenses, including per diem in lieu of subsistence, as authorized by section 5703 of title 5, United States Code. A member of the Commission not presently employed for compensation shall be compensated at a rate equal to the daily equivalent of the rate for GS-16 of the General Schedule under section 5332 of title 5, United States Code, for each day the member is engaged in the actual performance of his duties as a member of the Commission, not to exceed 90 days of service each year. Except for the purposes of chapter 81 of title 5 (relating to compensation for work injuries) and chapter 171 of title 28 (relating to tort claims), a member of the Commission shall not be considered an employee of the United States for any purpose.

(2) The Commission shall meet at the call of its Chairman or a majority of its members.

(3) Each Federal agency referred to in section 107(b) may designate a representative to participate as an observer with the Commission.

These representatives shall report to and advise the Commission on the activities relating to Arctic research of their agencies.

(4) The Commission shall conduct at least one public meeting in the State of Alaska annually.

DUTIES OF COMMISSION

SEC. 104. (a) The Commission shall-

(1) develop and recommend an integrated national Arctic research policy;

(2) in cooperation with the Interagency Arctic Research Policy Committee established under section 107, assist in establishing a national Arctic research program plan to implement the Arctic research policy;

(3) facilitate cooperation between the Federal Government and State and local governments with respect to Arctic research;

(4) review Federal research programs in the Arctic and suggest improvements in coordination among programs;

(5) recommend methods to improve logistical planning and support for Arctic research as may be appropriate and in accordance with the findings and purposes of this title;

(6) suggest methods for improving efficient sharing and dissemination of data and information on the Arctic among interested public and private institutions;

(7) offer other recommendations and advice to the Interagency Committee established under section 107 as it may find appropriate; and

(8) cooperate with the Governor of the State of Alaska and with agencies and organizations of that State which the Governor may designate with respect to the formulation of Arctic research policy.

(b) Not later than January 31 of each year, the Commission shall-

(1) publish a statement of goals and objectives with respect to Arctic research to guide the Interagency Committee established under section 107 in the performance of its duties; and

(2) submit to the President and to the Congress a report describing the activities and accomplishments of the Commission during the immediately preceding fiscal year.

COOPERATION WITH THE COMMISSION

SEC. 105. (a)(1) The Commission may acquire from the head of any Federal agency unclassified data, reports, and other nonproprietary information with respect to Arctic research in the possession of the agency which the Commission considers useful in the discharge of its duties.

(2) Each agency shall cooperate with the Commission and furnish all data, reports, and other information requested by the Commission to the extent permitted by law; except that no agency need furnish any information which it is permitted to withhold under section 552 of title 5, United States Code.

(b) With the consent of the appropriate agency head, the Commission may utilize the facilities and services of any Federal agency to the extent that the facilities and services are needed for the establishment and development of an Arctic research policy, upon reimbursement to be agreed upon by the Commission and the agency head and taking every feasible step to avoid duplication of effort.

(c) All Federal agencies shall consult with the Commission before undertaking major Federal actions relating to Arctic research.

ADMINISTRATION OF THE COMMISSION

SEC. 106. The Commission may-

(1) in accordance with the civil service laws and subchapter III of chapter 53 of title 5, United States Code, appoint and fix the compensation of an Executive Director and necessary additional staff personnel, but not to exceed a total of seven compensated personnel;

(2) procure temporary and intermittent services as authorized by section 3109 of title 5, United States Code;

(3) enter into contracts and procure supplies, services, and personal property; and

(4) enter into agreements with the General Services Administration for the procurement of necessary financial and administrative services, for which payment shall be made by reimbursement from funds of the Commission in amounts to be agreed upon by the Commission and the Administrator of the General Services Administration.

LEAD AGENCY AND INTERAGENCY ARCTIC RESEARCH POLICY COMMITTEE

SEC. 107. (a) The National Science Foundation is designated as the lead agency responsible for implementing Arctic research policy, and the Director of the National Science Foundation shall insure that the requirements of section 108 are fulfilled.

(b)(1) The President shall establish an Interagency Arctic Research Policy Committee (hereinafter referred to as the "Interagency Committee").

(2) The Interagency Committee shall be composed of representatives of the following Federal agencies or offices:

(A) the National Science Foundation;

(B) the Department of Commerce;

(C) the Department of Defense;

(D) the Department of Energy;

(E) the Department of the Interior;

(F) the Department of State;

(G) the Department of Transportation;

(H) the Department of Health and Human Services;

(I) the National Aeronautics and Space Administration;

(J) the Environmental Protection Agency; and

(K) any other agency or office deemed appropriate.

(3) The representative of the National Science Foundation shall serve as the Chairperson of the Interagency Committee.

DUTIES OF THE INTERAGENCY COMMITTEE

SEC. 108. (a) The Interagency Committee shall-

(1) survey Arctic research conducted by Federal, State, and local agencies, universities, and other public and private institutions to help determine priorities for future Arctic research, including natural resources and materials, physical and biological sciences, and social and behavioral sciences;

(2) work with the Commission to develop and establish an integrated national Arctic research policy that will guide Federal agencies in developing and implementing their research programs in the Arctic;

(3) consult with the Commission on-

(A) the development of the national Arctic research policy and the 5-year plan implementing the policy;

(B) Arctic research programs of Federal agencies;

(C) recommendations of the Commission on future Arctic research; and

(D) guidelines for Federal agencies for awarding and administering Arctic research grants;

(4) develop a 5-year plan to implement the national policy, as provided for in section 109;

(5) provide the necessary coordination, data, and assistance for the preparation of a single integrated, coherent, and multiagency budget request for Arctic research as provided for in section 110;

(6) facilitate cooperation between the Federal Government and State and local governments in Arctic research, and recommend the undertaking of neglected areas of research in accordance with the findings and purposes of this title;

(7) coordinate and promote cooperative Arctic scientific research programs with other nations, subject to the foreign policy guidance of the Secretary of State;

(8) cooperate with the Governor of the State of Alaska in fulfilling its responsibilities under this title;

(9) promote Federal interagency coordination of all Arctic research activities, including- (A) logistical planning and coordination; and

(B) the sharing of data and information associated with Arctic research, subject to section 552 of title 5, United States Code; and

(10) provide public notice of its meetings and an opportunity for the public to participate in the development and implementation of national Arctic research policy.

(b) Not later than January 31, 1986, and biennially thereafter, the Interagency Committee shall submit to the Congress through the President, a brief, concise report containing-

(1) a statement of the activities and accomplishments of the Interagency Committee since its last report; and

(2) a description of the activities of the Commission, detailing with particularity the recommendations of the Commission with respect to Federal activities in Arctic research.

5-YEAR ARCTIC RESEARCH PLAN

SEC. 109. (a) The Interagency Committee, in consultation with the Commission, the Governor of the State of Alaska, the residents of the Arctic, the private sector, and public interest groups, shall prepare a comprehensive 5-year program plan (hereinafter referred to as the "Plan") for the overall Federal effort in Arctic research. The Plan shall be prepared and submitted to the President for transmittal to the Congress within one year after the enactment of this Act and shall be revised biennially thereafter.

(b) The Plan shall contain but need not be limited to the following elements:

(1) an assessment of national needs and problems regarding the Arctic and the research necessary to address those needs or problems;

(2) a statement of the goals and objectives of the Interagency Committee for national Arctic research;

(3) a detailed listing of all existing Federal programs relating to Arctic research, including the existing goals, funding levels for each of the 5 following fiscal years, and the funds currently being expended to conduct the programs;

(4) recommendations for necessary program changes and other proposals to meet the requirements of the policy and goals as set forth by the Commission and in the Plan as currently in effect; and

(5) a description of the actions taken by the Interagency Committee to coordinate the budget review process in order to ensure interagency coordination and cooperation in (A) carrying out Federal Arctic research programs, and (B) eliminating unnecessary duplication of effort among these programs.

COORDINATION AND REVIEW OF BUDGET REQUESTS

SEC. 110. (a) The Office of Science and Technology Policy shall-

(1) review all agency and department budget requests related to the Arctic transmitted pursuant to section 108(a)(5), in accordance with the national Arctic research policy and the 5-year program under section 108(a)(2) and section 109, respectively; and

(2) consult closely with the Interagency Committee and the Commission to guide the Office of Science and Technology Policy's efforts.

(b)(1) The Office of Management and Budget shall consider all Federal agency requests for research related to the Arctic as one integrated, coherent, and multiagency request which shall be reviewed by the Office of Management and Budget prior to submission of the President's annual budget request for its adherence to the Plan. The Commission shall, after submission of the President's annual budget request, review the request and report to Congress on adherence to the Plan.

(2) The Office of Management and Budget shall seek to facilitate planning for the design, procurement, maintenance, deployment, and operations of icebreakers needed to provide a platform for Arctic research by allocating all funds necessary to support icebreaking operations, except for recurring incremental costs associated with specific projects, to the Coast Guard.

AUTHORIZATION OF APPROPRIATIONS; NEW SPENDING AUTHORITY

SEC. 111. (a) There are authorized to be appropriated such sums as may be necessary for carrying out this title.

(b) Any new spending authority (within the meaning of section 401 of the Congressional Budget Act of 1974) which is provided under this title shall be effective for any fiscal year only to such extent or in such amounts as may be provided in appropriation Acts.

DEFINITION

SEC. 112. As used in this title, the term "Arctic" means all United States and foreign territory north of the Arctic Circle and all United States territory north and west of the boundary formed by the Porcupine, Yukon, and Kuskokwim Rivers; all contiguous seas, including the Arctic Ocean and the Beaufort, Bering, and Chukchi Seas; and the Aleutian chain.

APPENDIX B. P.L. 101-609 OF 1990, AMENDING ARCTIC RESEARCH AND POLICY ACT (ARPA) OF 1984

The Arctic Research and Policy Act (ARPA) of 1984 (see **Appendix A**) was amended by P.L. 101-609 of November 16, 1990. The text of P.L. 101-609 is as follows:

SECTION 1. Except as specifically provided in this Act, whenever in this Act an amendment or repeal is expressed as an amendment to, or repeal of a provision, the reference shall be deemed to be made to the Arctic Research and Policy Act of 1984.

SEC. 2. Section 103(b)(1) (15 U.S.C. 4102(b)(1)) is amended—

(1) in the text above clause (A), by striking out `five' and inserting in lieu thereof `seven';

(2) in clause (A), by striking out `three' and inserting in lieu thereof `four'; and

(3) in clause (C), by striking out `one member' and inserting in lieu thereof `two members'.

SEC. 3. Section 103(d)(1) (15 U.S.C. 4102(d)(1)) is amended by striking out `GS-16' and inserting in lieu thereof `GS-18'.

SEC. 4. (a) Section 104(a) (15 U.S.C. 4102(a)) is amended—

(1) in paragraph (4), by striking out `suggest' and inserting in lieu thereof `recommend';

(2) in paragraph (6), by striking out `suggest' and inserting in lieu thereof `recommend';

(3) in paragraph (7), by striking out `and' at the end thereof;

(4) in paragraph (8), by striking out the period and inserting in lieu thereof a semicolon; and

(5) by adding at the end thereof the following new paragraphs:

'(9) recommend to the Interagency Committee the means for developing international scientific cooperation in the Arctic; and

'(10) not later than January 31, 1991, and every 2 years thereafter, publish a statement of goals and objectives with respect to Arctic research to guide the Interagency Committee established under section 107 in the performance of its duties.'.

(b) Section 104(b) is amended to read as follows:

'(b) Not later than January 31 of each year, the Commission shall submit to the President and to the Congress a report describing the activities and accomplishments of the Commission during the immediately preceding fiscal year.'.

SEC. 5. Section 106 (15 U.S.C. 4105) is amended—

(1) in paragraph (3), by striking out `and' at the end thereof;

(2) in paragraph (4), by striking out the period at the end thereof and inserting in lieu thereof; and'; and

(3) by adding at the end thereof the following new paragraph:

'(5) appoint, and accept without compensation the services of, scientists and engineering specialists to be advisors to the Commission. Each advisor may be allowed travel expenses, including per diem in lieu of subsistence, as authorized by section 5703 of title 5, United States Code. Except for the purposes of chapter 81 of title 5 (relating to compensation for work injuries) and chapter 171 of title 28 (relating to tort claims) of the United States Code, an advisor appointed under this paragraph shall not be considered an employee of the United States for any purpose.'

SEC. 6. Subsection (b)(2) of section 108 (15 U.S.C. 4107(b)(2)) is amended to read as follows:

'(2) a statement detailing with particularity the recommendations of the Commission with respect to Federal interagency activities in Arctic research and the disposition and responses to those recommendations.'

APPENDIX C. JANUARY 2009 ARCTIC POLICY DIRECTIVE (NSPD 66/HSPD 25)

On January 12, 2009, the George W. Bush Administration released a presidential directive establishing a new U.S. policy for the Arctic region. The directive, dated January 9, 2009, was issued as National Security Presidential Directive 66/Homeland Security Presidential Directive 25 (NSPD 66/HSPD 25). The text of NSPD 66/HSPD 25 is as follows:

SUBJECT: Arctic Region Policy

I. PURPOSE

A. This directive establishes the policy of the United States with respect to the Arctic region and directs related implementation actions. This directive supersedes Presidential Decision Directive/NSC-26 (PDD-26; issued 1994) with respect to Arctic policy but not Antarctic policy; PDD-26 remains in effect for Antarctic policy only.

B. This directive shall be implemented in a manner consistent with the Constitution and laws of the United States, with the obligations of the United States under the treaties and other international agreements to which the United States is a party, and with customary international law as recognized by the United States, including with respect to the law of the sea.

II. BACKGROUND

A. The United States is an Arctic nation, with varied and compelling interests in that region. This directive takes into account several developments, including, among others:

1. Altered national policies on homeland security and defense;

2. The effects of climate change and increasing human activity in the Arctic region;

3. The establishment and ongoing work of the Arctic Council; and

4. A growing awareness that the Arctic region is both fragile and rich in resources.

III. POLICY

A. It is the policy of the United States to:

1. Meet national security and homeland security needs relevant to the Arctic region;

2. Protect the Arctic environment and conserve its biological resources;

3. Ensure that natural resource management and economic development in the region are environmentally sustainable;

4. Strengthen institutions for cooperation among the eight Arctic nations (the United States, Canada, Denmark, Finland, Iceland, Norway, the Russian Federation, and Sweden);

5. Involve the Arctic's indigenous communities in decisions that affect them; and

6. Enhance scientific monitoring and research into local, regional, and global environmental issues.

B. National Security and Homeland Security Interests in the Arctic

1. The United States has broad and fundamental national security interests in the Arctic region and is prepared to operate either independently or in conjunction with other states to safeguard these interests. These interests include such matters as missile defense and early warning; deployment of sea and air systems for strategic sealift, strategic deterrence, maritime presence, and maritime security operations; and ensuring freedom of navigation and overflight.

2. The United States also has fundamental homeland security interests in preventing terrorist attacks and mitigating those criminal or hostile acts that could increase the United States vulnerability to terrorism in the Arctic region.

3. The Arctic region is primarily a maritime domain; as such, existing policies and authorities relating to maritime areas continue to apply, including those relating to law enforcement.[1] Human activity in the Arctic region is increasing and is projected to increase further in coming years. This requires the United States to assert a more active and influential national presence to protect its Arctic interests and to project sea power throughout the region.

4. The United States exercises authority in accordance with lawful claims of United States sovereignty, sovereign rights, and jurisdiction in the Arctic region, including sovereignty within the territorial sea, sovereign rights and jurisdiction within the United States exclusive economic zone and on the continental shelf, and appropriate control in the United States contiguous zone.

5. Freedom of the seas is a top national priority. The Northwest Passage is a strait used for international navigation, and the Northern Sea Route includes straits used for international navigation; the regime of transit passage applies to passage through those straits. Preserving the rights and duties relating to navigation and overflight in the Arctic region supports our ability to exercise these rights throughout the world, including through strategic straits.

6. Implementation: In carrying out this policy as it relates to national security and homeland security interests in the Arctic, the Secretaries of State, Defense, and Homeland Security, in coordination with heads of other relevant executive departments and agencies, shall:

a. Develop greater capabilities and capacity, as necessary, to protect United States air, land, and sea borders in the Arctic region;

b. Increase Arctic maritime domain awareness in order to protect maritime commerce, critical infrastructure, and key resources;

c. Preserve the global mobility of United States military and civilian vessels and aircraft throughout the Arctic region;

d. Project a sovereign United States maritime presence in the Arctic in support of essential United States interests; and

e. Encourage the peaceful resolution of disputes in the Arctic region.

C. International Governance

1. The United States participates in a variety of fora, international organizations, and bilateral contacts that promote United States interests in the Arctic. These include the Arctic Council, the International Maritime Organization (IMO), wildlife conservation and management agreements, and many other mechanisms. As the Arctic changes and human activity in the region increases, the United States and other governments should consider,

as appropriate, new international arrangements or enhancements to existing arrangements.

2. The Arctic Council has produced positive results for the United States by working within its limited mandate of environmental protection and sustainable development. Its subsidiary bodies, with help from many United States agencies, have developed and undertaken projects on a wide range of topics. The Council also provides a beneficial venue for interaction with indigenous groups. It is the position of the United States that the Arctic Council should remain a high-level forum devoted to issues within its current mandate and not be transformed into a formal international organization, particularly one with assessed contributions. The United States is nevertheless open to updating the structure of the Council, including consolidation of, or making operational changes to, its subsidiary bodies, to the extent such changes can clearly improve the Council's work and are consistent with the general mandate of the Council.

3. The geopolitical circumstances of the Arctic region differ sufficiently from those of the Antarctic region such that an "Arctic Treaty" of broad scope—along the lines of the Antarctic Treaty—is not appropriate or necessary.

4. The Senate should act favorably on U.S. accession to the U.N. Convention on the Law of the Sea promptly, to protect and advance U.S. interests, including with respect to the Arctic. Joining will serve the national security interests of the United States, including the maritime mobility of our Armed Forces worldwide. It will secure U.S. sovereign rights over extensive marine areas, including the valuable natural resources they contain. Accession will promote U.S. interests in the environmental health of the oceans. And it will give the United States a seat at the table when the rights that are vital to our interests are debated and interpreted.

5. Implementation: In carrying out this policy as it relates to international governance, the Secretary of State, in coordination with heads of other relevant executive departments and agencies, shall:

a. Continue to cooperate with other countries on Arctic issues through the United Nations (U.N.) and its specialized agencies, as well as through treaties such as the U.N. Framework Convention on Climate Change, the Convention on International Trade in Endangered Species of Wild Fauna and Flora, the Convention on Long Range Transboundary Air Pollution and its protocols, and the Montreal Protocol on Substances that Deplete the Ozone Layer;

b. Consider, as appropriate, new or enhanced international arrangements for the Arctic to address issues likely to arise from expected increases in human activity in that region, including shipping, local development and subsistence, exploitation of living marine resources, development of energy and other resources, and tourism;

c. Review Arctic Council policy recommendations developed within the ambit of the Council's scientific reviews and ensure the policy recommendations are subject to review by Arctic governments; and

d. Continue to seek advice and consent of the United States Senate to accede to the 1982 Law of the Sea Convention.

D. Extended Continental Shelf and Boundary Issues

1. Defining with certainty the area of the Arctic seabed and subsoil in which the United States may exercise its sovereign rights over natural resources such as oil, natural gas, methane hydrates, minerals, and living marine species is critical to our national interests in energy security, resource management, and environmental protection. The most effective way to achieve international recognition and legal certainty for our extended continental shelf is through the procedure available to States Parties to the U.N. Convention on the Law of the Sea.

2. The United States and Canada have an unresolved boundary in the Beaufort Sea. United States policy recognizes a boundary in this area based on equidistance. The

United States recognizes that the boundary area may contain oil, natural gas, and other resources.

3. The United States and Russia are abiding by the terms of a maritime boundary treaty concluded in 1990, pending its entry into force. The United States is prepared to enter the agreement into force once ratified by the Russian Federation.

4. Implementation: In carrying out this policy as it relates to extended continental shelf and boundary issues, the Secretary of State, in coordination with heads of other relevant executive departments and agencies, shall:

a. Take all actions necessary to establish the outer limit of the continental shelf appertaining to the United States, in the Arctic and in other regions, to the fullest extent permitted under international law;

b. Consider the conservation and management of natural resources during the process of delimiting the extended continental shelf; and

c. Continue to urge the Russian Federation to ratify the 1990 United States-Russia maritime boundary agreement.

E. Promoting International Scientific Cooperation

1. Scientific research is vital for the promotion of United States interests in the Arctic region. Successful conduct of U.S. research in the Arctic region requires access throughout the Arctic Ocean and to terrestrial sites, as well as viable international mechanisms for sharing access to research platforms and timely exchange of samples, data, and analyses. Better coordination with the Russian Federation, facilitating access to its domain, is particularly important.

2. The United States promotes the sharing of Arctic research platforms with other countries in support of collaborative research that advances fundamental understanding of the Arctic region in general and potential Arctic change in particular. This could include collaboration with bodies such as the Nordic Council and the European Polar Consortium, as well as with individual nations.

3. Accurate prediction of future environmental and climate change on a regional basis, and the delivery of near real-time information to end-users, requires obtaining, analyzing, and disseminating accurate data from the entire Arctic region, including both paleoclimatic data and observational data. The United States has made significant investments in the infrastructure needed to collect environmental data in the Arctic region, including the establishment of portions of an Arctic circumpolar observing network through a partnership among United States agencies, academic collaborators, and Arctic residents. The United States promotes active involvement of all Arctic nations in these efforts in order to advance scientific understanding that could provide the basis for assessing future impacts and proposed response strategies.

4. United States platforms capable of supporting forefront research in the Arctic Ocean, including portions expected to be ice-covered for the foreseeable future, as well as seasonally ice-free regions, should work with those of other nations through the establishment of an Arctic circumpolar observing network. All Arctic nations are members of the Group on Earth Observations partnership, which provides a framework for organizing an international approach to environmental observations in the region. In addition, the United States recognizes that academic and research institutions are vital partners in promoting and conducting Arctic research.

5. Implementation: In carrying out this policy as it relates to promoting scientific international cooperation, the Secretaries of State, the Interior, and Commerce and the Director of the National Science Foundation, in coordination with heads of other relevant executive departments and agencies, shall:

a. Continue to play a leadership role in research throughout the Arctic region;

b. Actively promote full and appropriate access by scientists to Arctic research sites through bilateral and multilateral measures and by other means;

c. Lead the effort to establish an effective Arctic circumpolar observing network with broad partnership from other relevant nations;

d. Promote regular meetings of Arctic science ministers or research council heads to share information concerning scientific research opportunities and to improve coordination of international Arctic research programs;

e. Work with the Interagency Arctic Research Policy Committee (IARPC) to promote research that is strategically linked to U.S. policies articulated in this directive, with input from the Arctic Research Commission; and

f. Strengthen partnerships with academic and research institutions and build upon the relationships these institutions have with their counterparts in other nations.

F. Maritime Transportation in the Arctic Region

1. The United States priorities for maritime transportation in the Arctic region are:

a. To facilitate safe, secure, and reliable navigation;

b. To protect maritime commerce; and

c. To protect the environment.

2. Safe, secure, and environmentally sound maritime commerce in the Arctic region depends on infrastructure to support shipping activity, search and rescue capabilities, short- and long-range aids to navigation, high-risk area vessel-traffic management, iceberg warnings and other sea ice information, effective shipping standards, and measures to protect the marine environment. In addition, effective search and rescue in the Arctic will require local, State, Federal, tribal, commercial, volunteer, scientific, and multinational cooperation.

3. Working through the International Maritime Organization (IMO), the United States promotes strengthening existing measures and, as necessary, developing new measures to improve the safety and security of maritime transportation, as well as to protect the marine environment in the Arctic region. These measures may include ship routing and reporting systems, such as traffic separation and vessel traffic management schemes in Arctic chokepoints; updating and strengthening of the Guidelines for Ships Operating in Arctic Ice-Covered Waters; underwater noise standards for commercial shipping; a review of shipping insurance issues; oil and other hazardous material pollution response agreements; and environmental standards.

4. Implementation: In carrying out this policy as it relates to maritime transportation in the Arctic region, the Secretaries of State, Defense, Transportation, Commerce, and Homeland Security, in coordination with heads of other relevant executive departments and agencies, shall:

a. Develop additional measures, in cooperation with other nations, to address issues that are likely to arise from expected increases in shipping into, out of, and through the Arctic region;

b. Commensurate with the level of human activity in the region, establish a risk-based capability to address hazards in the Arctic environment. Such efforts shall advance work on pollution prevention and response standards; determine basing and logistics support requirements, including necessary airlift and icebreaking capabilities; and improve plans and cooperative agreements for search and rescue;

c. Develop Arctic waterways management regimes in accordance with accepted international standards, including vessel traffic-monitoring and routing; safe navigation standards; accurate and standardized charts; and accurate and timely environmental and navigational information; and

d. Evaluate the feasibility of using access through the Arctic for strategic sealift and humanitarian aid and disaster relief.

G. Economic Issues, Including Energy

1. Sustainable development in the Arctic region poses particular challenges. Stakeholder input will inform key decisions as the United States seeks to promote economic and energy security. Climate change and other factors are significantly affecting the lives of Arctic inhabitants, particularly indigenous communities. The United States affirms the importance to Arctic communities of adapting to climate change, given their particular vulnerabilities.

2. Energy development in the Arctic region will play an important role in meeting growing global energy demand as the area is thought to contain a substantial portion of the world's undiscovered energy resources. The United States seeks to ensure that energy development throughout the Arctic occurs in an environmentally sound manner, taking into account the interests of indigenous and local communities, as well as open and transparent market principles. The United States seeks to balance access to, and development of, energy and other natural resources with the protection of the Arctic environment by ensuring that continental shelf resources are managed in a responsible manner and by continuing to work closely with other Arctic nations.

3. The United States recognizes the value and effectiveness of existing fora, such as the Arctic Council, the International Regulators Forum, and the International Standards Organization.

4. Implementation: In carrying out this policy as it relates to economic issues, including energy, the Secretaries of State, the Interior, Commerce, and Energy, in coordination with heads of other relevant executive departments and agencies, shall:

a. Seek to increase efforts, including those in the Arctic Council, to study changing climate conditions, with a view to preserving and enhancing economic opportunity in the Arctic region. Such efforts shall include inventories and assessments of villages, indigenous communities, subsistence opportunities, public facilities, infrastructure, oil and gas development projects, alternative energy development opportunities, forestry, cultural and other sites, living marine resources, and other elements of the Arctic's socioeconomic composition;

b. Work with other Arctic nations to ensure that hydrocarbon and other development in the Arctic region is carried out in accordance with accepted best practices and internationally recognized standards and the 2006 Group of Eight (G-8) Global Energy Security Principles;

c. Consult with other Arctic nations to discuss issues related to exploration, production, environmental and socioeconomic impacts, including drilling conduct, facility sharing, the sharing of environmental data, impact assessments, compatible monitoring programs, and reservoir management in areas with potentially shared resources;

d. Protect United States interests with respect to hydrocarbon reservoirs that may overlap boundaries to mitigate adverse environmental and economic consequences related to their development;

e. Identify opportunities for international cooperation on methane hydrate issues, North Slope hydrology, and other matters;

f. Explore whether there is a need for additional fora for informing decisions on hydrocarbon leasing, exploration, development, production, and transportation, as well as shared support activities, including infrastructure projects; and

g. Continue to emphasize cooperative mechanisms with nations operating in the region to address shared concerns, recognizing that most known Arctic oil and gas resources are located outside of United States jurisdiction.

H. Environmental Protection and Conservation of Natural Resources

1. The Arctic environment is unique and changing. Increased human activity is expected to bring additional stressors to the Arctic environment, with potentially serious consequences for Arctic communities and ecosystems.

2. Despite a growing body of research, the Arctic environment remains poorly understood. Sea ice and glaciers are in retreat. Permafrost is thawing and coasts are eroding.

Pollutants from within and outside the Arctic are contaminating the region. Basic data are lacking in many fields. High levels of uncertainty remain concerning the effects of climate change and increased human activity in the Arctic. Given the need for decisions to be based on sound scientific and socioeconomic information, Arctic environmental research, monitoring, and vulnerability assessments are top priorities. For example, an understanding of the probable consequences of global climate variability and change on Arctic ecosystems is essential to guide the effective long-term management of Arctic natural resources and to address socioeconomic impacts of changing patterns in the use of natural resources.

3. Taking into account the limitations in existing data, United States efforts to protect the Arctic environment and to conserve its natural resources must be risk-based and proceed on the basis of the best available information.

4. The United States supports the application in the Arctic region of the general principles of international fisheries management outlined in the 1995 Agreement for the Implementation of the Provisions of the United Nations Convention on the Law of the Sea of December 10, 1982, relating to the Conservation and Management of Straddling Fish Stocks and Highly Migratory Fish Stocks and similar instruments. The United States endorses the protection of vulnerable marine ecosystems in the Arctic from destructive fishing practices and seeks to ensure an adequate enforcement presence to safeguard Arctic living marine resources.

5. With temperature increases in the Arctic region, contaminants currently locked in the ice and soils will be released into the air, water, and land. This trend, along with increased human activity within and below the Arctic, will result in increased introduction of contaminants into the Arctic, including both persistent pollutants (e.g., persistent organic pollutants and mercury) and airborne pollutants (e.g., soot).

6. Implementation: In carrying out this policy as it relates to environmental protection and conservation of natural resources, the Secretaries of State, the Interior, Commerce, and Homeland Security and the Administrator of the Environmental Protection Agency, in coordination with heads of other relevant executive departments and agencies, shall:

a. In cooperation with other nations, respond effectively to increased pollutants and other environmental challenges;

b. Continue to identify ways to conserve, protect, and sustainably manage Arctic species and ensure adequate enforcement presence to safeguard living marine resources, taking account of the changing ranges or distribution of some species in the Arctic. For species whose range includes areas both within and beyond United States jurisdiction, the United States shall continue to collaborate with other governments to ensure effective conservation and management;

c. Seek to develop ways to address changing and expanding commercial fisheries in the Arctic, including through consideration of international agreements or organizations to govern future Arctic fisheries;

d. Pursue marine ecosystem-based management in the Arctic; and

e. Intensify efforts to develop scientific information on the adverse effects of pollutants on human health and the environment and work with other nations to reduce the introduction of key pollutants into the Arctic.

IV. Resources and Assets

A. Implementing a number of the policy elements directed above will require appropriate resources and assets. These elements shall be implemented consistent with applicable law and authorities of agencies, or heads of agencies, vested by law, and subject to the availability of appropriations. The heads of executive departments and agencies with responsibilities relating to the Arctic region shall work to identify future budget, administrative, personnel, or legislative proposal requirements to implement the elements of this directive.

[1] These policies and authorities include Freedom of Navigation (PDD/NSC-32), the U.S. Policy on Protecting the Ocean Environment (PDD/NSC-36), Maritime Security Policy (NSPD-41/HSPD-13), and the National Strategy for Maritime Security (NSMS).[334]

APPENDIX D. MAY 2013 NATIONAL STRATEGY FOR ARCTIC REGION

On May 10, 2013, the Obama Administration released a document entitled *National Strategy for the Arctic Region.*[335] The executive summary of the document is reprinted earlier in this report (see "May 2013 National Strategy for Arctic Region" in "Background"). This appendix reprints the main text of the document. The main text states:

Introduction

We seek an Arctic region that is stable and free of conflict, where nations act responsibly in a spirit of trust and cooperation, and where economic and energy resources are developed in a sustainable manner that also respects the fragile environment and the interests and cultures of indigenous peoples.

As the United States addresses these opportunities and challenges, we will be guided by our central interests in the Arctic region, which include providing for the security of the United States; protecting the free flow of resources and commerce; protecting the environment; addressing the needs of indigenous communities; and enabling scientific research. In protecting these interests, we draw from our long-standing policy and approach to the global maritime spaces in the 20th century, including freedom of navigation and overflight and other internationally lawful uses of the sea and airspace related to these freedoms; security on the oceans; maintaining strong relationships with allies and partners; and peaceful resolution of disputes without coercion.

To achieve this vision, the United States is establishing an overarching national approach to advance national security interests, pursue responsible stewardship of this precious and unique region, and serve as a basis for cooperation with other Arctic states[336] and the international community as a whole to advance common interests.

Even as we work domestically and internationally to minimize the effects of climate change, the effects are already apparent in the Arctic. Ocean resources are more readily accessible as sea ice diminishes, but thawing ground is threatening communities as well as hindering land-based activities, including access to resources. Diminishing land and sea ice is altering ecosystems and the services they provide. As an Arctic nation, the United States must be proactive and disciplined in addressing changing regional conditions and in developing adaptive strategies to protect its interests. An undisciplined approach to exploring new opportunities in this frontier could result in significant harm to the region, to our national security interests, and to the global good.

When implementing this strategy, the United States will proceed in a thoughtful, responsible manner that leverages expertise, resources, and cooperation from the State of

Alaska, Alaska Natives, and stakeholders across the entire nation and throughout the international community. We will encourage and use science-informed decisionmaking to aid this effort.

We will endeavor to do no harm to the sensitive environment or to Alaska native communities and other indigenous populations that rely on Arctic resources. Just as a common spirit and shared vision of peaceful partnership led to the development of an international space station, we believe much can be achieved in the Arctic region through collaborative international efforts, coordinated investments, and public-private partnerships.

Structure of the Strategy

Through this National Strategy for the Arctic Region, we seek to guide, prioritize, and synchronize efforts to protect U.S. national and homeland security interests, promote responsible stewardship, and foster international cooperation.

This strategy articulates three priority lines of effort. It also identifies guiding principles as a foundation for Arctic region activities. Through a deliberate emphasis on the priority lines of effort and objectives, it aims to achieve a national unity of effort that is consistent with our domestic and international legal rights, obligations, and commitments and that is well coordinated with our Arctic neighbors and the international community. These lines of effort identify common themes where specific emphasis and activities will be focused to ensure that strategic priorities are met. The three lines of effort, as well as the guiding principles are meant to be acted upon as a coherent whole.

Changing Conditions

While the Arctic region has experienced warming and cooling cycles over millennia, the current warming trend is unlike anything previously recorded. The reduction in sea ice has been dramatic, abrupt, and unrelenting. The dense, multi-year ice is giving way to thin layers of seasonal ice, making more of the region navigable year-round. Scientific estimates of technically recoverable conventional oil and gas resources north of the Arctic Circle total approximately 13 percent of the world's undiscovered oil and 30 percent of the world's undiscovered gas deposits, as well as vast quantities of mineral resources, including rare earth elements, iron ore, and nickel.[337] These estimates have inspired fresh ideas for commercial initiatives and infrastructure development in the region. As portions of the Arctic Ocean become more navigable, there is increasing interest in the viability of the Northern Sea Route and other potential routes, including the Northwest Passage, as well as in development of Arctic resources.

For all of the opportunities emerging with the increasing accessibility and economic and strategic interests in the Arctic, the opening and rapid development of the Arctic region presents very real challenges. On the environmental front, reduced sea ice is having an immediate impact on indigenous populations as well as on fish and wildlife. Moreover, there may be potentially profound environmental consequences of continued ocean warming and Arctic ice melt. These consequences include altering the climate of lower latitudes, risking the stability of Greenland's ice sheet, and accelerating the thawing of the Arctic permafrost in which large quantities of methane – a potent driver of climate change – as well as pollutants such as mercury are stored. Uncoordinated development – and the consequent increase in pollution such as emissions of black carbon or other substances from fossil fuel combustion – could have unintended consequences on climate trends, fragile ecosystems, and Arctic communities. It is imperative that the United States proactively establish national priorities and objectives for the Arctic region.

Lines of Effort

To meet the challenges and opportunities in the Arctic region, and in furtherance of established Arctic Region Policy,[338] we will pursue the following lines of effort and supporting objectives in a mutually reinforcing manner that incorporates the broad range of U.S. current activities and interests in the Arctic region.

1. Advance United States Security Interests

Our highest priority is to protect the American people, our sovereign territory and rights, natural resources, and interests of the United States. To this end, the United States will identify, develop, and maintain the capacity and capabilities necessary to promote safety, security, and stability in the region through a combination of independent action, bilateral initiatives, and multilateral cooperation. We acknowledge that the protection of our national security interests in the Arctic region must be undertaken with attention to environmental, cultural, and international considerations outlined throughout this strategy. As many nations across the world aspire to expand their role in the Arctic, we encourage Arctic and non-Arctic states to work collaboratively through appropriate fora to address the emerging challenges and opportunities in the Arctic region, while we remain vigilant to protect the security interests of the United States and our allies.

To accomplish this line of effort, the United States Government will seek to:

- **Evolve Arctic Infrastructure and Strategic Capabilities** – Working cooperatively with the State of Alaska, local, and tribal authorities, as well as public and private sector partners, we will develop, maintain, and exercise the capacity to execute Federal responsibilities in our Arctic waters, airspace, and coastal regions, including the capacity to respond to natural or man-made disasters. We will carefully tailor this regional infrastructure, as well as our response capacity, to the evolving human and commercial activity in the Arctic region.
- **Enhance Arctic Domain Awareness** – We seek to improve our awareness of activities, conditions, and trends in the Arctic region that may affect our safety, security, environmental, or commercial interests. The United States will endeavor to appropriately enhance sea, air, and space capabilities as Arctic conditions change, and to promote maritime-related information sharing with international, public, and private sector partners, to support implementation of activities such as the search-and-rescue agreement signed by Arctic states.
- **Preserve Arctic Region Freedom of the Seas** – The United States has a national interest in preserving all of the rights, freedoms, and uses of the sea and airspace recognized under international law. We will enable prosperity and safe transit by developing and maintaining sea, under-sea, and air assets and necessary infrastructure. In addition, the United States will support the enhancement of national defense, law enforcement, navigation safety, marine environment response, and search-and-rescue capabilities. Existing international law provides a comprehensive set of rules governing the rights, freedoms, and uses of the world's oceans and airspace, including the Arctic. The law recognizes these rights, freedoms, and uses for commercial and military vessels and aircraft. Within this framework, we shall further develop Arctic waterways management regimes, including traffic separation schemes, vessel tracking, and ship routing, in collaboration with partners. We will also encourage other nations to adhere to internationally accepted principles. This cooperation will facilitate strategic partnerships that promote innovative, low-cost solutions that enhance the Arctic marine transportation system and the safe, secure, efficient and free flow of trade.

- **Provide for Future United States Energy Security** – The Arctic region's energy resources factor into a core component of our national security strategy: energy security. The region holds sizable proved and potential oil and natural gas resources that will likely continue to provide valuable supplies to meet U.S. energy needs. Continuing to responsibly develop Arctic oil and gas resources aligns with the United States "all of the above" approach to developing new domestic energy sources, including renewables, expanding oil and gas production, and increasing efficiency and conservation efforts to reduce our reliance on imported oil and strengthen our nation's energy security. Within the context of this broader energy security strategy, including our economic, environmental and climate policy objectives, we are committed to working with stakeholders, industry, and other Arctic states to explore the energy resource base, develop and implement best practices, and share experiences to enable the environmentally responsible production of oil and natural gas as well as renewable energy.

2. Pursue Responsible Arctic Region Stewardship

Responsible stewardship requires active conservation of resources, balanced management, and the application of scientific and traditional knowledge of physical and living environments. As Arctic environments change, increased human activity demands precaution, as well as greater knowledge to inform responsible decisions. Together, Arctic nations can responsibly meet new demands – including maintaining open sea lanes for global commerce and scientific research, charting and mapping, providing search-and-rescue services, and developing capabilities to prevent, contain, and respond to oil spills and accidents – by increasing knowledge and integrating Arctic management.[339] We must improve our ability to forecast future conditions in the Arctic while being mindful of the potential for unexpected developments.

To realize this line of effort, we will pursue the specific objectives outlined below:

- **Protect the Arctic Environment and Conserve Arctic Natural Resources** – Protecting the unique and changing environment of the Arctic is a central goal of U.S. policy. Supporting actions will promote healthy, sustainable, and resilient ecosystems over the long term, supporting a full range of ecosystem services. This effort will be risk-based and proceed on the basis of best available information. The United States in the Arctic will assess and monitor the status of ecosystems and the risks of climate change and other stressors to prepare for and respond effectively to environmental challenges.
- **Use Integrated Arctic Management to Balance Economic Development, Environmental Protection, and Cultural Values** – Natural resource management will be based on a comprehensive understanding of environmental and cultural sensitivities in the region, and address expectations for future infrastructure needs and other development-related trends. This endeavor can promote unity of effort and provide the basis for sensible infrastructure and other resource management decisions in the Arctic. We will emphasize science-informed decisionmaking and integration of economic, environmental, and cultural values. We will also advance coordination among Federal departments and agencies and collaboration with partners engaged in Arctic stewardship activities.
- **Increase Understanding of the Arctic through Scientific Research and Traditional Knowledge** – Proper stewardship of the Arctic requires understanding of how the environment is changing, and such understanding will be based on a holistic earth system approach. Vast areas of the Arctic Ocean are

unexplored, and we lack much of the basic knowledge necessary to understand and address Arctic issues. The changes in the Arctic cannot be understood in isolation and must be viewed in a global context. As we learn more about the region, we have identified several key subcomponents of the Arctic that require urgent attention: land ice and its role in changing sea level; sea-ice and its role in global climate, fostering biodiversity, and supporting Arctic peoples; and, the warming permafrost and its effects on infrastructure and climate. Better earth system-level knowledge will also help us meet operational needs such as weather and ice forecasting. We can make faster progress through a well-coordinated and transparent national and international exploration and research agenda that reduces the potential for duplication of effort and leads to better leveraging of resources.

- **Chart the Arctic region** – We will continue to make progress in charting and mapping the Arctic region's ocean and waterways, so long obscured by perennial ice, and mapping its coastal and interior lands according to reliable, modern standards. Given the vast expanse of territory and water to be charted and mapped, we will need to prioritize and synchronize charting efforts to make more effective use of resources and attain faster progress. In so doing, we will make navigation safer and contribute to the identification of ecologically sensitive areas and reserves of natural resources.

3. Strengthen International Cooperation

What happens in one part of the Arctic region can have significant implications for the interests of other Arctic states and the international community as a whole. The remote and complex operating conditions in the Arctic environment make the region well-suited for collaborative efforts by nations seeking to explore emerging opportunities while emphasizing ecological awareness and preservation. We will seek to strengthen partnerships through existing multilateral fora and legal frameworks dedicated to common Arctic issues. We will also pursue new arrangements for cooperating on issues of mutual interest or concern and addressing unique and unprecedented challenges, as appropriate.

U.S. efforts to strengthen international cooperation and partnerships will be pursued through four objectives:

- **Pursue Arrangements that Promote Shared Arctic State Prosperity, Protect the Arctic Environment, and Enhance Security** – We will seek opportunities to pursue efficient and effective joint ventures, based on shared values that leverage each Arctic state's strengths. This collaboration will assist in guiding investments and regional activities, addressing dynamic trends, and promoting sustainable development in the Arctic region.

  Arctic nations have varied commercial, cultural, environmental, safety, and security concerns in the Arctic region. Nevertheless, our common interests make these nations ideal partners in the region. We seek new opportunities to advance our interests by proactive engagement with other Arctic nations through bilateral and multilateral efforts using of a wide array of existing multilateral mechanisms that have responsibilities relating to the Arctic region.

  As appropriate, we will work with other Arctic nations to develop new coordination mechanisms to keep the Arctic region prosperous, environmentally sustainable, operationally safe, secure, and free of conflict, and will protect U.S., allied, and regional security and economic interests.
- **Work through the Arctic Council to Advance U.S. Interests in the Arctic Region** – In recent years, the Arctic Council has facilitated notable

achievements in the promotion of cooperation, coordination, and interaction among Arctic states and Arctic indigenous peoples. Recent successes of the Council include its advancement of public safety and environmental protection issues, as evidenced by the 2011 Arctic Search-and-Rescue Agreement and by the 2013 Arctic Marine Oil Pollution Preparedness and Response Agreement. The United States will continue to emphasize the Arctic Council as a forum for facilitating Arctic states' cooperation on myriad issues of mutual interest within its current mandate.

- **Accede to the Law of the Sea Convention** – Accession to the Convention would protect U.S. rights, freedoms, and uses of the sea and airspace throughout the Arctic region, and strengthen our arguments for freedom of navigation and overflight through the Northwest Passage and the Northern Sea Route. The United States is the only Arctic state that is not party to the Convention. Only by joining the Convention can we maximize legal certainty and best secure international recognition of our sovereign rights with respect to the U.S. extended continental shelf in the Arctic and elsewhere, which may hold vast oil, gas, and other resources. Our extended continental shelf claim in the Arctic region could extend more than 600 nautical miles from the north coast of Alaska.
  In instances where the maritime zones of coastal nations overlap, Arctic states have already begun the process of negotiating and concluding maritime boundary agreements, consistent with the Law of the Sea Convention and other relevant international law. The United States supports peaceful management and resolution of disputes, in a manner free from coercion. While the United States is not currently a party to the Convention, we will continue to support and observe principles of established customary international law reflected in the Convention.
- **Cooperate with other Interested Parties** – A growing number of non-Arctic states and numerous non-state actors have expressed increased interest in the Arctic region. The United States and other Arctic nations should seek to work with other states and entities to advance common objectives in the Arctic region in a manner that protects Arctic states' national interests and resources. One key example relates to the promotion of safe, secure, and reliable Arctic shipping, a goal that is best pursued through the International Maritime Organization in coordination with other Arctic states, major shipping states, the shipping industry and other relevant interests.

Guiding Principles

The U.S. approach to the Arctic region must reflect our values as a nation and as a member of the global community. We will approach holistically our interests in promoting safety and security, advancing economic and energy development, protecting the environment, addressing climate change and respecting the needs of indigenous communities and Arctic state interests. To guide our efforts, we have identified the following principles to serve as the foundation for U.S. Arctic engagement and activities.

- **Safeguard Peace and Stability** by working to maintain and preserve the Arctic region as an area free of conflict, acting in concert with allies, partners, and other interested parties. This principle will include United States action, and the actions of other interested countries, in supporting and preserving international legal principles of freedom of navigation and overflight and other uses of the sea related to these freedoms, unimpeded lawful commerce, and the peaceful resolution of disputes. The United States will rely on existing international law, which provides a comprehensive set of rules governing the rights, freedoms, and uses of the world's oceans and airspace, including the Arctic.

- **Make Decisions Using the Best Available Information** by promptly sharing – nationally and internationally – the most current understanding and forecasts based on up-to-date science and traditional knowledge.
- **Pursue Innovative Arrangements** to support the investments in scientific research, marine transportation infrastructure requirements, and other support capability and capacity needs in this region. The harshness of the Arctic climate and the complexity associated with developing, maintaining, and operating infrastructure and capabilities in the region necessitate new thinking on public-private and multinational partnerships.
- **Consult and Coordinate with Alaska Natives** consistent with tribal consultation policy established by Executive Order.[340] This policy emphasizes trust, respect, and shared responsibility. It articulates that tribal governments have a unique legal relationship with the United States and requires Federal departments and agencies to provide for meaningful and timely input by tribal officials in development of regulatory policies that have tribal implications. This guiding principle is also consistent with the Alaska Federation of Natives Guidelines for Research.[341]

Conclusion

We seek a collaborative and innovative approach to manage a rapidly changing region. We must advance U.S. national security interests, pursue responsible stewardship, and strengthen international collaboration and cooperation, as we work to meet the challenges of rapid climate-driven environmental change. The melting of Arctic ice has the potential to transform global climate and ecosystems as well as global shipping, energy markets, and other commercial interests. To address these challenges and opportunities, we will align Federal activities in accordance with this strategy; partner with the State of Alaska, local, and tribal entities; and work with other Arctic nations to develop complementary approaches to shared challenges. We will proactively coordinate regional development. Our economic development and environmental stewardship must go hand-in-hand. The unique Arctic environment will require a commitment by the United States to make judicious, coordinated infrastructure investment decisions, informed by science. To meet this challenge, we will need bold, innovative thinking that embraces and generates new and creative public-private and multinational cooperative models.[342]

End Notes

[1] Except for the subsection on the Arctic and the U.N. Convention on the Law of the Sea, this section was prepared by Ronald O'Rourke, Specialist in Naval Affairs, Foreign Affairs, Defense, and Trade Division.

[2] On November 25, 2008, voters in Greenland approved a referendum for greater autonomy that some observers view as a step toward eventual independence from Denmark. (Alan Cowell, "Greenland Vote Favors Independence," *New York Times*, November 26, 2008.)

[3] For more on the Arctic Council on the Internet, see http://www.arctic-council.org/.

[4] Title II of P.L. 98-373 is the National Critical Materials Act of 1984.

[5] As codified, the definition reads, "As used in this chapter.... "

[6] For discussions and (in some cases) maps, see Susan Joy Hassol, *Impacts of a Warming Arctic* [Executive Summary]. Cambridge, Cambridge University Press, 2004, p. 4, available at http://amap.no/acia/; Oran R. Yong and Niels Einarsson, *Arctic Human Development Report*, Stefansson Arctic Institute, Akureyri, Iceland, 2004, pp. 17-18, available at http://www.svs.is/AHDR/AHDR%20chapters/English%20version/Chapters%20PDF.htm; and Hugo Ahlenius, editor in chief et al. *Vital Arctic Graphics, People and Global Heritage on*

Our Last Wild Shores, UNEP/GRID-Arendal, Arendal, Norway, p. 6, available at http://www.grida.no/publications/vg/arctic/.

[7] A map showing the line that results from 10° isotherm definition is available at https://www.cia.gov/library/publications/the-world-factbook/reference_maps/pdf/arctic.pdf.

[8] Discussion entitled "Geographical Coverage," available at http://www.amap.no/ (click on "About AMAP" and then the tab "Geographical coverage.")

[9] Discussion entitled "Geographical Coverage," available at http://www.amap.no/ (click on "About AMAP" and then the tab "Geographical coverage.")

[10] Title II of P.L. 98-373 is the National Critical Materials Act of 1984.

[11] These words are taken from the official title of P.L. 98-373. (Arctic Research and Policy Act of 1984 is the short title of Title I of P.L. 98-373.) The remainder of P.L. 98-373's official title relates to Title II of the act, the short title of which is the National Critical Materials Act of 1984.)

[12] The IARPC currently includes more than a dozen federal agencies, departments, and offices. Additional information on the IARPC is available at http://www.nsf.gov/od/opp/arctic/iarpc/start.jsp.

[13] NSF used to conduct its Arctic research activities through a Division of Arctic Sciences (ARC) that formed part of an Office of Polar Programs (OPP). NSF states that in FY2013, a realignment of offices previously under the Office of the Director was implemented, and that as a part of this realignment, OPP became PLR within GEO. (National Science Foundation, *FY 2014 Budget Request to Congress*, April 10, 2013, Footnote 2 in Summary Tables—3 (pdf page 25 of 506). The document is available at http://www.nsf.gov/about/budget/fy2014/pdf/EntireDocument_fy2014.pdf.)

[14] National Science Foundation, *FY 2015 Budget Request to Congress*, March 10, 2014, p. GEO-16 (pdf page 128 of 510). Accessed March 28, 2014, at http://www.nsf.gov/about/budget/fy2015/pdf/Entire_ Document_ FY2015.pdf.

[15] National Science Foundation, *FY 2015 Budget Request to Congress*, March 10, 2014, p. GEO-1 (pdf page 113 of 510).

[16] SEES stands for Science, Engineering, and Education for Sustainability.

[17] The passage at this point contains a footnote that provides the following Internet address: http://www.nsf.gov/od/opp/arctic/iarpc/arc_res_plan_index.jsp.

[18] The passage at this point contains a footnote that provides the following Internet address: http://www.whitehouse.gov/administration/eop/oceans

[19] National Science Foundation, *FY 2015Budget Request to Congress*, March 10, pp. GEO-16 and GEO-17 (pdf pages 128 and 129 of 510).

[20] National Science Foundation, *FY 2015 Budget Request to Congress*, March 10, 2014, p. GEO-17 (pdf page 129 of 510).

[21] CRS communication with State Department official, October 8, 2010.

[22] *National Security Strategy*, Washington, May 2010, p. 50. The quoted sentence constitutes the entirety of the document's comments specifically on the Arctic. It is the final sentence of a section on "sustain[ing] broad cooperation on key global challenges" that includes longer discussions on climate change, peacekeeping and armed conflict, pandemics and infectious disease, transnational criminal threats and threats to governance, and safeguarding the global commons.

[23] *National Strategy for the Arctic Region*, May 2013, 11 pp.; accessed May 14, 2013, at http://www.whitehouse.gov/ sites/default/files/docs/nat_arctic_strategy.pdf. The document includes a cover letter from President Obama dated May 10, 2013.

[24] *National Strategy for the Arctic Region* states on page 6 that the "lines of effort" it describes are to be undertaken "[t]o meet the challenges and opportunities in the Arctic region, and in furtherance of established Arctic Region Policy," at which point there is a footnote referencing the January 2009 Arctic policy directive.

[25] A footnote in the document at this point states: "Traditional knowledge refers to a body of evolving practical knowledge based on observations and personal experience of indigenous communities over an extensive, multigenerational time period. (BOEM Ocean Science, Vol. 9, Issue 2, May/April/June 2012, page 4)."

[26] *National Strategy for the Arctic Region*, May 2013, pp. 2-3.

[27] *Implementation Plan for The National Strategy for the Arctic Region*, January 2014, 32 pp. The news release anouncing the implementation plan is posted at http://www.whitehouse.gov/blog/2014/01/30/white-house-releasesimplementation-plan-national-strategy-arctic-region. The document itself is posted at http://www.whitehouse.gov/sites/
default/files/docs/implementation_plan_for_the_national_strategy_for_the_arctic_region_-_fi.... pdf.

[28] *Implementation Plan for The National Strategy for the Arctic Region*, January 2014, p. 1.

[29] The passage contains a footnote at this point stating that this executive order was signed by the President on November 1, 2013. It is Executive Order 13653. The text of the order is posted at http://www.gpo.gov/fdsys/pkg/FR2013-11-06/pdf/2013-26785.pdf and http://www.whitehouse.gov/the-press-office/2013/11/01/executive-orderpreparing-united-states-. A fact sheet about it is posted at http://www.whitehouse.gov/the-pressoffice/2013/11/01/fact-sheet-executive-order-climate-preparedness

[30] *Implementation Plan for The National Strategy for the Arctic Region*, January 2014, p. 4.

[31] This section prepared by Marjorie Ann Browne, Specialist in International Relations, Foreign Affairs, Defense, and Trade Division.

[32] The United States is party to the four conventions adopted in 1958: Convention on the Territorial Sea and the Contiguous Zone, Convention on the High Seas, Convention on the Continental Shelf, and Convention on Fishing and Conservation of the Living Resources of the High Seas.

[33] Treaty Document 103-39.

[34] Other relevant provisions of the Convention, applicable depending on the extent of Arctic melting, relate to navigation, high seas freedoms, fisheries, and exclusive economic zones.

[35] The continental shelf is the under-sea extension of a coastal state's land territory. Article 76 of the Convention defines the continental shelf, *inter alia*, as "the seabed and subsoil of the submarine areas that extend beyond its [coastal state's] territorial sea throughout the natural prolongation of its land territory to the outer edge of the continental margin."

[36] A coastal State party has 10 years from the entry into force of the Convention for submission of information on its proposed limits. In May 2001, the Meeting of States Parties to the Convention decided that for any State for which the Convention entered into force before May 13, 1999, the date of commencement of the 10-year time period for making submissions to the commission is May 13, 1999.

[37] Annex II, Article 9. Article 83 of the Convention provides that questions relating to these boundary delimitation disputes shall be resolved by agreement between the States or by the Dispute Settlement options set forth in Part XV of the Convention.

[38] Article 76, para. 8.

[39] For more information, see http://www.state

[40] *Implementation Plan for The National Strategy for the Arctic Region*, January 2014, p. 29.

[41] *Implementation Plan for The National Strategy for the Arctic Region*, January 2014, p. 29.

[42] This section prepared by Jane Leggett, Specialist in Energy and Environmental Policy, Resources, Science, and Industry Division.

[43] In scientific analyses, "ice-free" does not necessarily mean "no ice." The definition of "ice-free" or sea ice "extent" or "area" varies across studies. Sea ice "extent" is one common measure, equal to the sum of the area of grid cells that have ice concentration of less than a set percentage—frequently 15%. For more information, see the National Snow and Ice Data Center, http://nsidc.org/seaice/data/terminology.html.

[44] Muyin Wang and James E. Overland, "A Sea Ice Free Summer Arctic within 30 Years?," *Geophysical Research Letters* 36, no. L07502 (April 3, 2009): 10.1029/2009GL037820; Marika Holland, Cecilia M. Bitz, and Bruno Tremblay, "Future abrupt reductions in the summer Arctic sea ice," *Geophysical Research Letters* 33, no. L23503 (2006); But see also Julien Boé, Alex Hall, and Xin Qu, "Sources of spread in simulations of Arctic sea ice loss over the twenty-first century," *Climatic Change* 99, no. 3 (April 1, 2010): 637-645; I. Eisenman and J. S. Wettlaufer, "Nonlinear threshold behavior during the loss of Arctic sea ice," *Proceedings of the National Academy of Sciences* 106, no. 1 (January 6, 2009): 28-32; Dirk Notz, "The Future of Ice Sheets and Sea Ice: Between Reversible Retreat and Unstoppable Loss," *Proceedings of the National Academy of Sciences* 106, no. 49 (December 8, 2009): 20590-20595.

[45] Overland et al. state that "a warm Arctic-cold continent pattern represents a paradox of recent global warming: there is not a uniform pattern of temperature increases" due to a set of newly recognized processes described in Overland, J. E, K. R Wood, and M. Wang. "Warm Arctic-cold Continents: Climate Impacts of the Newly Open Arctic Sea." *Polar Research* 30 (2011). The authors raise a critical, unanswered question, "Is the observed severe mid-latitude weather in two adjacent years simply due to an extreme in chaotic processes alone, or do they included a partial but important Arctic forcing and connection due to recent changing conditions?" In other words, are recent patterns random anomalies, or might we expect more of the same?; among other examples, see also Lim, Young-Kwon, and Siegfried D. Schubert. "The Impact of ENSO and the Arctic Oscillation on Winter Temperature Extremes in the Southeast United States." *Geophysical Research Letters* 38, no. 15 (August 11, 2011): L15706.

[46] There was a regionally warm period in the Arctic from the mid-1920s to around 1940, which scientists have assessed to have been driven by natural climate variability. They have found that period to be distinctly

different from the recent multi-decadal warming, in part because the early 20th century warming was concentrated in the northern high latitudes. See, for example, Figure 2, upper left graphic, in Geophysical Fluid Dynamics Laboratory, "Simulatoin of Early 20th Century Warming," at http://www.gfdl.noaa.gov/early-20th-century-global-warming.

[47] Steele, Michael, Wendy Ermold, and Jinlun Zhang. "Arctic Ocean Surface Warming Trends over the Past 100 Years." *Geophysical Research Letters* 35, no. 2 (January 29, 2008): L02614.

[48] J. Overland, J. Walsh, and M. Wang, *Arctic Report Card - Atmosphere* (NOAA Arctic Research Program, October 6, 2008), http://www.arctic.noaa.gov/reportcard/atmosphere

[49] National Snow and Ice Data Center, "Arctic sea ice extent breaks 2007 record low" (August 27, 2012); Japanese Aerospace Exploration Agency, "A new record minimum of the Arctic sea ice extent was set on 24 August 2012"; Arctic ROOS (Norway), "Daily Updated Time series of Arctic sea ice area and extent derived from SSMI data provided by NERSC," at http://arctic-roos.org/observations/satellite-data/sea-ice/ice-area-and-extent-in-arctic.

[50] William L. Chapman and John E. Walsh, "Simulations of Arctic Temperature and Pressure by Global Coupled Models," *Journal of Climate* 20, no. 4 (February 1, 2007): 609-632.

[51] See footnote 43. Also, although one Canadian scientist has predicted that recurrent ice-free summers may begin sometime between 2013 and 2020, this is not consistent with other climate models' projections.

[52] This section was prepared by Carl Ek, who was a Specialist in International Relations, Foreign Affairs, Defense, and Trade Division, until his retirement on April 30, 2014. For questions relating to this section, contact Derek E. Mix, Analyst in European Affairs, Foreign Affairs, Defense, and Trade Division.

[53] "Russia, Canada Make Competing Claims To Arctic Resources," *The Canadian Press*, September 16, 2010.

[54] "5 Countries Agree To Talk, Not Compete, Over the Arctic," *New York Times*, May 29, 2008.

[55] For additional information, see Natalie Mychajlyszyn, *The Arctic: Geopolitical Issues*, Canadian Library of Parliament, October 24, 2008.

[56] "Russia, Norway Sign Deal On Barents Sea Border, Seek More Development in Mineral-rich Arctic," *Associated Press*, September 15, 2010.

[57] Cannon quoted in "Canada Seeks To Settle Arctic Borders," *Agence France Presse*, August 20, 2010. For additional information concerning Canada's August statement on Arctic policy, see "Statement on Canada's Arctic foreign policy: Exercising sovereignty and promoting Canada's Northern Strategy abroad," Foreign Affairs and International Trade Canada website, http://www.international.gc.ca/polar-polaire/canada_arctic_foreign_policyla_politique_etrangere_du_canada_pour_arctique.aspx?lang=eng.

[58] "Canada and Denmark Sign Arctic Cooperation Arrangement," *Targeted News Service*, May 17, 2010.

[59] This section prepared by John Frittelli, Specialist in Transportation Policy, Resources, Science, and Industry Division.

[60] Extended daylight hours in the Arctic during the summer may also be an advantage.

[61] Presentation by Stephen Carmel, Senior Vice President, Maersk Line Ltd., Halifax International Security Forum, *Arctic Security: The New Great Game?* November 21, 2009, available at http://fora.tv/.

[62] A third but more remote possibility is a route directly over the North Pole.

[63] Northern Sea Route Information Office; http://www.arctic-lio.com/

[64] This was the route pioneered by the *SS Manhattan*, an oil tanker modified for ice breaking in 1969 to carry Alaskan North Slope oil to the Atlantic. This was the first commercial passage through the NWP, but the building of the Alaskan pipeline was found to be the more economical means of transporting oil from the North Slope to the lower 48 states.

[65] Although the NWP is often compared to the alternative route through the Panama Canal in terms of distance and sailing days from Asia to the U.S. east coast, another alternative to consider is the shorter and faster transcontinental rail route across Canada or the United States. The Panama Canal is undergoing an expansion project, expected to be completed by 2015, to allow larger ships with about three times the cargo capacity to pass through.

[66] " Cargo Ship Carves a Path in Arctic Sea," *The Wall Street Journal*, September 26, 2013, p. A11.

[67] "U.S. Needs to Prepare for Arctic Traffic Surge," *Anchorage Daily News*, February 12, 2008, p. A4.

[68] "Northern Waters," *The Globe and Mail* (Canada), August 14, 2008, p. A8.

[69] See http://www.aeco.no/index.htm for more information.

[70] S.E.L. Howell and J.J. Yackel, "A Vessel Transit Assessment of Sea Ice Variability in the Western Arctic, 1969-2002: Implications for Ship Navigation," *Canadian Journal of Remote Sensing*, vol. 30, no. 2, 2004.

[71] "Arctic Unlikely to See Major Shipping Growth," *New Zealand Transport and Logistics Business Week*, April 24, 2008.

[72] Stephen M. Carmel, Senior Vice President, Maersk Line Ltd., "The Cold, Hard Realities of Arctic Shipping," *United States Naval Institute, Proceedings*; July 2013, pp. 38-41.

[73] In July and August 2010, NOAA surveyed the Bering Straits area in order to update its charts but stated that it will take more than 25 years to map the prioritized areas of navigational significance in U.S. Arctic waters. See http://www.noaanews.noaa.gov/stories2010/20100720_fairweather.html.

[74] Ice reporting that currently exists is intended for scientists not mariners.

[75] For further information, see http://www.poa.usace.army.mil/en/cw/AKPortsStudy.htm, and FY2013 USACE Budget Justification, p. POD-5.

[76] *Alaska Deep-Draft Arctic Port System Study,* March 2013; http://www.poa.usace.army.mil/Library/ReportsandStudies/AlaskaRegionalPortsStudy.aspx. The navigation channel at Nome presently ranges from 10 to 20 feet in depth. Much of the harbor at Port Clarence has a natural depth of 35 to 40 feet; http://www.charts.noaa.gov/ OnLineViewer/AlaskaViewerTable.shtml.

[77] U.S. Committee on the Marine Transportation System, *U.S. Arctic Marine Transportation System: Overview and Priorities for Action*, 2013; http://www.cmts.gov/downloads/CMTS_Arctic_MTS

[78] See http://www.imo.org/ for more information.

[79] See MSC/Circ. 1056/MEPC/Circ.399, at http://www.imo.org/.

[80] Edwin H. Anderson, "Polar Shipping, The Forthcoming Polar Code and Implications for the Polar Environments," *Journal of Maritime Law and Commerce,* v. 43, no. 1, January 2012.

[81] The Coast Guard is studying shipping routes through the Bering Strait for possible safety enhancements. See 75 FR 68568, November 8, 2010.

[82] This section prepared by Curry Hagerty, Specialist in Energy and Natural Resources Policy, Resources, Science, and Industry Division; Peter Folger, Specialist in Energy and Natural Resources Policy, Resources, Science, and Industry Division; and Marc Humphries, Analyst in Energy Policy, Resources, Science, and Industry Division.

[83] Data provided by written communication to CRS from Bureau of Safety and Environmental Enforcement on October 4-11, 2012. There are no federally regulated production facilities or pipelines in the Alaska region.

[84] One way of delineating the extent of the continental margins that is currently under way in the region is by the terms of Article 76 of the U.N. Convention on the Law of the Sea (UNCLOS).

[85] CRS Report R41132, *Outer Continental Shelf Moratoria on Oil and Gas Development*, by Curry L. Hagerty.

[86] FY2004 DOI Appropriations (P.L. 108-108). Furthermore, the Continuing Appropriations Resolution 2009 (P.L. 110-329) did not extend the annual congressional moratorium on oil and gas leasing activities in the lower 48 states. On March 11, 2009, the Omnibus Appropriations Act, 2009 (P.L. 111-8) was enacted without moratorium provisions, confirming that the oil and gas development bans in federal waters along the Atlantic and Pacific coasts, parts of Alaska, and the Gulf of Mexico that had been in place since 1982 had not been restored in 2009 appropriations measures.

[87] "*Withdrawal of Certain Areas of the United States Outer Continental Shelf from Leasing Disposition*," §12(a) OCSLA, 43 U.S.C. 1341(a) (March 31, 2010). This memorandum is available at http://www.doi.gov/whatwedo/energy ocs/upload/2010alaska-mem-rel.pdf.

[88] See statement of the Alaska OCS Region at http://www.boem.gov/About-BOEM/BOEM-Regions/Alaska-Region/ Index.aspx. On March 27, 2012, BOEM issued a Request for Interest, with respect to the Cook Inlet planning area.

[89] For example with respect to pre-leasing activities in the Cook Inlet, on March 27, 2012, BOEM issued a Request for Interest, with respect to the Cook Inlet planning area. In light of responses to the Request, BOEM decided to proceed with the pre-sale process for the Cook Inlet and to place the date for a potential lease sale in 2016 to allow time to complete the necessary steps to develop additional resource and environmental information, and conduct analysis required by NEPA.

[90] For statements of legislative interest, see House Committee on Natural Resources 113th Congress Oversight Plan (adopted January 23, 2013). For statements of interest by President Obama, see 2013 State of the Union address at http://www.whitehouse.gov/state

[91] 43 U.S.C. §1344. The current Five Year Program took effect August 27, 2012, and is expected to remain in effect until 2017. 77 *Federal Register* 40080 (July 6, 2012).

[92] On April 20, 2010, an explosion occurred at the *Deepwater Horizon* drilling platform in the Gulf of Mexico, resulting in 11 fatalities. The platform sank into the Gulf on April 22, 2010, amid a significant release of oil at the sea floor. According to the National Incident Command's Flow Rate Technical Group estimate, the well released approximately 206 million gallons (4.9 million barrels) of oil before it was contained July 15, 2010.

[93] Shell is a global group of energy and petrochemical companies with 90,000 employees in more than 80 countries. In the United States, Shell operates in 50 states as an oil and natural gas producer, gasoline and natural gas marketer, and petrochemical manufacturer. See more at http://www.shell.com.

[94] Among other approvals, on August 4, 2011, BOEM issued a conditional approval of Shell's revised exploration plan (EP) for the Beaufort Sea to drill up to four exploration wells starting in summer 2012. On December 16, 2011, BOEM issued conditional approval of Shell's revised EP in the Chukchi Sea to drill up to six exploration wells beginning in the 2012 drilling season.

[95] Shell's Kulluk ran aground off the southern coast of Alaska in a storm. The rig was towed to safe harbor, with no serious injuries to the crew.

[96] The U.S. Coast Guard conducted this review with the Bureau of Safety and Environmental Enforcement (BSEE) and the National Transportation Safety Board providing technical assistance. For more details on this review see "*Secretary Salazar Launches Expedited Assessment of 2012 Arctic Operations High-Level, Expedited Review of the 2012 Offshore Drilling Program in the Beaufort and Chukchi Seas Will Assess Activities, Provide Lessons Learned*," (January 8, 2013), at http://www.doi.gov/news/pressreleases/secretary-salazar-launches-expedited-assessment-

[97] For Mr. Luthi's full statement see *Keep Cool Head on Arctic Energy Development*, by Amy Harder, National Journal, reporting comments submitted by Randall Luthi, President of the National Ocean Industries Association (January 24, 2013),available at http://energy

[98] For a full discussion of this perspective see *We're Not Yet Arctic-Ready* by Marilyn Heiman, Director of the Pew Environment Group's U.S. Arctic Program (January 21, 2013). This statement is available at http://energy

[99] Prior to DOI approvals issued in 2012, Shell had attempted to establish an exploratory drilling program offshore Alaska in the Beaufort and Chukchi Seas for several years. Shell's long-standing goal has been to confirm a major discovery of oil in commercially-viable quantities in the Alaskan Arctic Ocean. For a comprehensive discussion of case law featuring Shell as a defendant and as a plaintiff in litigation over federal leasing in the Alaska program areas see CRS Report WSLG332, *A Preemptive Legal Strike? Shell Seeks Injunction to Prevent Environmental Group Challenges to Drilling Permits*, by Adam Vann.

[100] DOI required Shell to include in its drilling plan an in-theater subsea containment system capable of functioning in the event of an emergency and subject to review and oversight by federal agencies responsible for Arctic exploration. For further information see "*Secretary Salazar Launches Expedited Assessment of 2012 Arctic Operations*," January 8, 2013, available at http://www.doi.gov/news/pressreleases/secretary-salazar-launches-expedited-assessment arctic-operations.cfm.

[101] See Report to the Secretary of the Interior; Review of Shell's 2012 Alaska Offshore Oil and Gas Exploration Program (March 8, 2013), at http://www.doi.gov/news/pressreleases/upload/Shell-report-3-8-13-Final.pdf.

[102] A seismic array is typically a long string or streamer of geophones—acoustic devices used for recording seismic signals—towed behind a ship while the ship traverses a prospective oil and gas-bearing portion of the seafloor. The seismic signals are processed and interpreted to give a cross-section or three-dimensional image of the subsurface.

[103] See USGS Circum-Arctic Resource Appraisal website at http://energy

[104] USGS Fact Sheet 2008-3049: Circum-Arctic Resource Appraisal: Estimates of Undiscovered Oil and Gas North of the Arctic Circle, http://pubs.usgs.gov/fs/2008/3049/.

[105] Donald L. Gautier et al., "Assessment of Undiscovered Oil and Gas in the Arctic," *Science*, vol. 324, no. 5931, pp. 1175-1179 (2009).

[106] The Associated Press, "Greenland Opens to Oil Firms; Melting Ice Unlocks Reserves," International Herald Tribune (January 14, 2008), at http://www.iht.com/articles/ap/2008/01/14/business/NA-FIN-US-Greenland-Oil-Companies.php.

[107] *Secretary Salazar Unveils Arctic Studies Initiative that will Inform Oil and Gas Decisions for Beaufort and Chukchi Seas*: Press Release April 13, 2010, http://www.doi.gov/news/pressreleases/2010_04_13_releaseA.cfm.

[108] U.S. Geological Survey Fact Sheet 2011-3048, 4 p. (2011).

[109] Tony Halpin, "President Medvedev Threatens Russian Arctic Annexation," Times Online (September 18, 2008), at http://www.timesonline.co.uk/tol/news/world/europe/article4773567.ece.

[110] The purpose of the U.S. Extended Continental Shelf Project is to establish the full extent of the continental shelf of the United States, consistent with international law. Involved in this mission are the U.S. Coast Guard (USCG), Department of State (DOS), and the University of New Hampshire (UNH). NOAA has the lead in collecting bathymetric data. The U.S. Geological Survey has the lead in collecting seismic data.

[111] See USGS, U.S. Extended Continental Shelf Project, at http://www.state

[112] A prominent directive issued by the Bush Administration addressed, among other issues, national security and maritime transportation. National Security Presidential Directive/NSPD-66 and Homeland Security Presidential Directive/HSPD-25, at http://www.whitehouse.gov/news/releases/2009/01/20090112-3.html.

[113] This initiative is part of an effort Arctic Research Commission Chairman Fran Ulmer began in December 2011. For a full discussion of this initiative see "*Obama Administration Outlines Arctic Energy Policy Initiatives*," Oil and Gas Journal (February 7, 2012), available at http://www.ogj.com/articles/2012/02/obama-administration-outlines-arcticenergy-policy

[114] Holland-Bartels, Leslie, and Pierce, Brenda, eds., 2011, An evaluation of the science needs to inform decisions on Outer Continental Shelf energy development in the Chukchi and Beaufort Seas, Alaska: U.S. Geological Survey Circular 1370, 278 p. (2011). This report is available at http://pubs.usgs.gov/circ/1370/.

[115] Don Gautier, Research Geologist, USGS, quoted in "Countries in Tug-of-War Over Arctic Resources," CNN (January 2, 2009), at http://www.cnn.com/2009/TECH/science/01/02/arctic.rights.dispute/index.html?iref=newssearch.

[116] Carolyn Fitzpatrick, "Heavy Haul in the High North," Railway Gazette International (July 24, 2008), at http://www.railwaygazette.com/news/single-view/view/10/heavy-haul-in-the-high-north.html.

[117] Cecelia Jamasmie, *Melting Arctic is nothing but good news for the mining industry: Lloyd's/Chatham House*, Mining.com, April 13, 2012, http://www.mining.com/melting-

[118] Charles Emmerson and Glada Lahn, *Arctic Opening: Opportunity and Risk in the High North*, Chatham House, 2012, http://www.lloyds.com/~/media Arctic_Risk_ Report_20120412.pdf.

[119] See National Oceanic and Atmospheric Administration, Arctic Change, at http://www.arctic.noaa.gov/detect /landroad.shtml?page=land.

[120] This section prepared by Jonathan L. Ramseur, Specialist in Environmental Policy, Resources, Science, and Industry Division.

[121] Arctic Monitoring and Assessment Programme (AMAP), Arctic Oil and Gas 2007 (2008).

[122] Arctic Monitoring and Assessment Programme (AMAP), Arctic Oil and Gas 2007 (2008).

[123] See this report's section "Implications for Sea Transportation," by John Fritelli.

[124] For example, the *Exxon Valdez* spilled approximately 11 million gallons of oil, but its carrying capacity was approximately 60 million gallons.

[125] See this report's section "Implication of Changes in the Arctic for Oil, Gas, and Mineral Exploration and Development," by Peter Folger and Marc Humphries.

[126] Larger oil spills occurred during the 1991 Iraq War, but many of those spills were deliberate. A 1910-1911 onshore oil blowout in the California San Joaquin Valley is reported to have spilled 9.4 million barrels of crude oil (almost 400 million gallons).

[127] See CRS Report R41531, *Deepwater Horizon Oil Spill: The Fate of the Oil*, by Jonathan L. Ramseur.

[128] National Research Council (NRC) of the National Academies of Science, *Oil in the Sea III: Inputs, Fates, and Effects* (2003).

[129] See CRS Report RL33705, *Oil Spills in U.S. Coastal Waters: Background and Governance*, by Jonathan L. Ramseur; and Dagmar Etkin (Environmental Research Consulting), Analysis of U.S. Oil Spillage, Prepared for American Petroleum Institute, August 2009.

[130] National Research Council of the National Academies of Science, *Cumulative Environmental Effects of Oil and Gas Activities on Alaska's North Slope* (2003).

[131] See, for example, Minerals Management Service (MMS), Outer Continental Shelf Oil & Gas Leasing Program: 2007-2012, Final Environmental Impact Statement, April 2007, Chapter 4; MMS, Proposed Gulf of Mexico OCS Oil and Gas Lease Sale 206, Central Planning Area, Environmental Assessment, October 2007.

[132] Letter from Marvin E. Odum, President, Shell Oil Company to S. Elizabeth Birnbaum, Minerals Management Service (May 14, 2010). Cited in a staff paper from the National Commission on the BP Deepwater Horizon Oil Spill and Offshore Drilling ("The Challenges of Oil Spill Response in the Arctic," January 2011).

[133] AMAP, *Arctic Oil and Gas 2007 (2008).*

[134] AMAP, *Arctic Oil and Gas 2007* (2008).

[135] World Wildlife Fund, *Oil Spill: Response Challenges in Arctic Waters* (2007).

[136] Coastal Response Research Center, *Opening the Arctic Seas: Envisioning Disasters and Framing Solutions* (2009), partnership between the National Oceanic and Atmospheric Administration and the University of New Hampshire.

[137] Although the response gap in the Arctic has not been quantified, a recent estimate of Prince William Sound (PWS) may be instructive. A 2007 study found a response gap for PWS of 38% for the time of the study period (65% during the winter season). Note that PWS has existing infrastructure for response, while the more

remote Arctic areas do not. Nuka Research and Planning Group, LLC, *Response Gap Estimate for Two Operating Areas in Prince William Sound, Alaska* (2007), Report to Prince William Sound Regional Citizens' Advisory Council.

[138] Arthur Brooks (Commander, 17th Coast Guard District) Coast Guard Journal, Arctic Journal (April 7, 2008).

[139] U.S. Coast Guard, *Report to Congress: U.S. Coast Guard Polar Operations*, December 2008.

[140] A single road connects Deadhorse, Alaska and the Prudhoe Bay with central Alaska (Fairbanks). For more details and maps of the area, see Nuka Research and Planning Group, *Oil Spill Prevention and Response in the U.S. Arctic Ocean: Unexamined Risks, Unacceptable Consequences*, Commissioned by Pew Environment Group, November 2010.

[141] Government Accountability Office, *Coast Guard: Efforts to Identify Arctic Requirements Are Ongoing, but More Communication about Agency Planning Efforts Would Be Beneficial*, GAO-10-870, September 2010.

[142] Transportation Research Board of the National Academy of Sciences, *Risk of Vessel Accidents and Spills in the Aleutian Islands: Designing a Comprehensive Risk Assessment* (2008), Special Report 293, National Academies Press. Washington, DC.

[143] Coastal Response Research Center, *Opening the Arctic Seas: Envisioning Disasters and Framing Solutions*, (2009), partnership between the National Oceanic and Atmospheric Administration and the University of New Hampshire. See also, U.S. Arctic Research Commission, (Draft) *White Paper: U.S. Arctic Research Commission Recommends Steps to Expanded U.S. Funding for Arctic/Subarctic Oil Spill Research* (February 24, 2010).

[144] Arctic Monitoring and Assessment Programme (AMAP), *Arctic Oil and Gas 2007* (2008).

[145] National Research Council, *Cumulative Environmental Effects of Oil and Gas Activities on Alaska's North Slope* (2003).

[146] Arturo A. Keller and Kristin Clark, *Oil Recovery with Novel Skimmer Surfaces under Cold Climate Conditions* (2007), prepared for the Minerals Management Service.

[147] World Wildlife Fund, *Oil Spill: Response Challenges in Arctic Waters* (2007). For further discussion of issues relating to oil spills, see CRS Report RL33705, *Oil Spills in U.S. Coastal Waters: Background and Governance*, by Jonathan L. Ramseur.

[148] Coastal Response Research Center, *Opening the Arctic Seas: Envisioning Disasters and Framing Solutions*, (2009), partnership between the National Oceanic and Atmospheric Administration and the University of New Hampshire.

[149] This section prepared by Eugene Buck, Specialist in Natural Resources Policy, Resources, Science, and Industry Division.

[150] Erik J. Molenaar and Robert Corell, *Arctic Fisheries*, Arctic Transform, February 9, 2009; available at http://arctictransform.org/download/FishBP.pdf.

[151] 74 *Federal Register* 56734-56746, November 3, 2009.

[152] Information on the symposium is available at http://www.nprb.org/iafs2009/.

[153] The website is available at http://alaskafisheries.noaa.gov/sustainablefisheries/arctic/.

[154] Prepared by Lynne Corn and Eugene Buck, Specialists in Natural Resources Policy, Resources, Science, and Industry Division.

[155] Fish and Wildlife Service, "Endangered and Threatened Wildlife and Plants; Review of Native Species That Are Candidates for Listing as Endangered or Threatened; Annual Notice of Findings on Resubmitted Petitions; Annual Description of Progress on Listing Actions; Proposed Rule," 76 *Federal Register* 66370-66439, October 26, 2001.

[156] 73 *Federal Register* 79822-79828.

[157] 75 *Federal Register* 65239-65248.

[158] 74 *Federal Register* 53683-58696, October 20, 2009.

[159] 75 *Federal Register* 77476-77495.

[160] 75 *Federal Register* 77496-77515.

[161] Among biologists, it is traditionally said that a species faced with extreme change can respond in three basic ways: "migrate, mutate, or die." When change is rapid enough, mutation (accompanied by natural selection of individuals within the population more suited to the changed environment) may not be able to occur fast enough, leaving migration and death as the only options. The problem of response rate is more severe for species that reproduce slowly (e.g., polar bears) and less severe for species that reproduce rapidly (e.g., algae).

[162] The efficacy and the effect of this tactic is often questioned, since natural migration is unlikely to involve the entire suite of species in an ecosystem (e.g., host plants might not move north (or up) as fast as their moth herbivores, nor as fast as the birds that depend on the moths). Moreover, the southerners will not find a land of

sterile bare dirt—the species that are already there may be threatened themselves by the competition from the new arrivals, perhaps tipping the balance and pushing still more species toward extinction.

[163] Under the MMPA, both NMFS and FWS have responsibility for additional marine mammal species (e.g., manatees, sea otters, dolphins) which are not currently found in the Arctic.

[164] This section was originally prepared by Roger Walke, who was a Specialist in American Indian Policy, Domestic Social Policy Division, until his retirement from CRS in October 2010. It has been updated by Jane A. Leggett, Specialist in Environmental and Energy Policy in CRS's Resources, Science and Industry Division.

[165] *Arctic Human Development Report*, ed. Joan Nymand Larsen et al. (Akureyri, Iceland: Stefansson Arctic Institute, 2004), p. 47; this report is subsequently cited in this section as *AHDR.* The seven countries are Canada, Denmark-Greenland, Finland, Norway, Russia, Sweden, and the United States.

[166] John F. Hoffecker, *A Prehistory of the North: Human Settlement of the Higher Latitudes* (New Brunswick, NJ: Rutgers University Press, 2005), pp. 8, 81, 112-115.

[167] *AHDR*, pp. 47, 53; David Crystal, *Cambridge Encyclopedia of Language*, 2nd ed. (Cambridge: Cambridge University Press, 1997), chap. 50; *Ethnologue: Languages of the World*, 16th ed., ed. M. Paul Lewis (Dallas: SIL International, 2009), available at http://www.ethnologue.com/. The number of languages and language families varies not only with definitions of the Arctic but with definitions of languages and language families.

[168] Jim Berner et al., *Arctic Climate Impact Assessment* (Cambridge: Cambridge University Press, 2005), chapter 12; this report is subsequently cited in this section as *ACIA*.

[169] *ACIA*, pp. 654-655.

[170] Fennoscandia refers to the Scandinavian Peninsula, Finland, the Kola Peninsula of Russia, and certain parts of Russia bordering on Finland.

[171] Janet Martin, *Treasure in the Land of Darkness: The Fur Trade and Its Significance for Medieval Russia* (Cambridge: Cambridge University Press, 1986), pp. 41-42; James Forsyth, *A History of the Peoples of Siberia: Russia's North Asian Colony, 1581-1990* (Cambridge: Cambridge University Press, 1992), pp. 69-83, 102; Lassi K. Heininen, "Different Images of the Arctic and the Circumpolar North in World Politics," in *Knowledge and Power in the Arctic*, Proceedings at a Conference in Rovaniemi, April 16-18, 2007, Arctic Centre Reports 48, ed. Paula Kankaanpaa et al. (Rovaniemi, Finland: University of Lapland, Arctic Centre, 2007), p. 125.

[172] James W. VanStone, "Exploration and Contact History of Western Alaska," and David Damas, "Copper Eskimo," and J. Garth Taylor, "Historical Ethnography of the Labrador Coast," in *Handbook of North American Indian: Vol. 5, Arctic,* vol. ed. David Damas, gen. ed. William C. Sturtevant (Washington: Smithsonian, 1984), pp. 149-155, 408, 509- 510.

[173] Inge Kleivan, "History of Norse Greenland," in *Handbook, Vol. 5, Arctic, op. cit.*, pp. 549-555; Finn Gad, "Danish Greenland Policies," in *Handbook of North American Indians: Vol. 4, History of Indian-White Relations*, vol. ed. Wilcomb E. Washburn, gen. ed. William C. Sturtevant (Washington: Smithsonian, 1988), p. 110.

[174] Parkinson, Alan J. The Arctic Human Health Initiative. Washington DC: U.S. Department of Health and Human Services, 2006.

[175] *AHDR*, pp. 19, 29. Estimates of Arctic indigenous populations are complicated by varying definitions not only of the Arctic but also of indigenous peoples; for instance, Russia does not count some non-European Arctic ethnic groups, such as the Yakut, as "indigenous minorities" (see "Peoples of the Arctic: Characteristics of Human Populations Relevant to Pollution Issues," in *AMAP Assessment Report: Arctic Pollution Issues*, ed. Simon J. Wilson et al. (Oslo: Arctic Monitoring and Assessment Programme, 1998), pp. 167-169; this report is subsequently cited in this section as *AMAP 1998*.

[176] *AMAP 1998*, chapter 5; see also Birger Poppel et al., *SLiCA Results*, Survey of Living Conditions in the Arctic (Anchorage: Institute of Social and Economic Research, University of Alaska Anchorage, 2007), pp. 4-7, http://www.arcticlivingconditions.org.

[177] Annika E. Nilson and Henry P. Huntington, *Arctic Pollution 2009* (Oslo: Arctic Monitoring and Assessment Programme, 2009), pp. 39-41; this report is subsequently cited in this section as *AMAP 2009.*

[178] *ACIA*, p. 1000.

[179] *SLiCA Results, op.cit.*, pp. v, 4-8.

[180] *AHDR*, p. 232.

[181] *AHDR*, chapter 4, and pp. 232-233.

[182] *AHDR*, chapters 6-7, and pp. 232-233.

[183] *ACIA*, pp. 1000-1001, 1004.

[184] *ACIA*, pp. 1000-1001, 1004.

[185] Parkinson, Alan J. *The Arctic Human Health Initiative*. Washington DC: U.S. Department of Health and Human Services, 2006; Brubaker, Michael, James Berner, Raj Chavan, and John Warren. "Climate Change and Health Effects in Northwest Alaska." *Global Health Action* 4 (October 18, 2011).

[186] U.S. Government Accountability Office (GAO), *Alaska Native Villages: Villages Affected by Flooding and Erosion Have Difficulty Qualifying for Federal Assistance*, GAO-04-895T, June 29, 2004, p. i, http://www.gao.gov/new.items/ d04895t.pdf. See also, Government Accountability Office, *Alaska Native Villages: Most Are Affected by Flooding and Erosion, but Few Qualify for Federal Assistance*, GAO-04-142, December 12, 2003, http://www.gao.gov/new.items/ d04142.pdf.

[187] U.S. Army Corps of Engineers. Alaska Baseline Erosion Assessment (BEA), March 2009.

[188] GAO, *Alaska Native Villages: Limited Progress Has Been Made on Relocating Villages Threatened by Flooding and Erosion*, June 3, 2009.

[189] Alaska Federation of Natives, Human Resources Committee, *2010 Federal Priorities* (Anchorage: Alaska Federation of Natives, 2010), pp. 22-23, available at http://www.nativefederation.org/documents/ 2010_AFN_Federal_Priorities.pdf. See also, Government Accountability Office, *Alaska Native Villages: Limited Progress Has Been Made on Relocating Villages Threatened by Flooding and Erosion*, GAO-09-551, June 3, 2009, http://www.gao.gov/new.items/d09551.pdf. For a more detailed anecdote, see a 2012 interview with Brice Eningowuk, Shishmaref Erosion and Relocation Committee member at Shearer, Christine. "Climate Crisis: Alaskan Village Shishmaref Sinking Into the Sea." *Imagined Magazine*, July 8, 2012, http://imaginedmag.com/2012/07/climate-

[190] Press Office of Senator Lisa Murkowski. "Murkowski Urges Greater Tribal Consultation from Administration; Senator Spotlights 'Monumental' Climate Struggles Faced by Alaska Natvies, Coastal Communities", July 19, 2012.

[191] *ACIA*, pp. 1001, 1004.

[192] Aqqaluk Lynge, Chair, Inuit Circumpolar Council, quoted in Stephanie McFeeters, "Lynge talks future of Inuit people," *The Dartmouth*, February 8, 2012.

[193] See, for example, "Health: Increased Bacterial Loads in Potable Water Could Have Significant Health Effects on Indigenous People From the Arctic to Uganda, Says Vanier Scholar." National Aboriginal Health Organization (NAHO), February 17, 2012. http://www.naho.ca/blog/2012/02/17/health- or, Kallenborn et al. *Combined Effects of Selected Pollutants and Climate Change in the Arctic Environment*. Oslo, Norway: Arctic Monitoring and Assessment Programme (AMAP), Arctic Council, 2011.

[194] *AMAP Assessment 2009: Human Health in the Arctic*, ed. Simon J. Wilson and Carolyn Symon (Oslo: Arctic Monitoring and Assessment Programme, 2009)), pp. 4-6, 143.

[195] Brubaker, Michael, James Berner, Raj Chavan, and John Warren. "Climate Change and Health Effects in Northwest Alaska." *Global Health Action* 4 (October 18, 2011); John Warren, "Climate change could affect human health," *Mukluk Telegraph*, January/February 2005, pp. 5-6.

[196] John Warren, "Climate change could affect human health," *Mukluk Telegraph*, January/February 2005, pp. 5-6.

[197] *AHDR*, p. 235.

[198] See http://www.arctic-council.org/. The six organizations are the Aleut International Association, Arctic Athabaskan Council, Gwich'in Council International, Inuit Circumpolar Council, RAIPON (Russian Association of Indigenous Peoples of the North), and Saami Council.

[199] United Nations, Declaration on the Rights of Indigenous Peoples, Resolution A/RES/61/295, at http://social.un.org/ index/IndigenousPeoples/DeclarationontheRightsofIndigenousPeoples.aspx.

[200] See http://www.indigenoussummit.com/servlet/content/home.html.

[201] K. Galloway-McLean et al., *Report of the Indigenous Peoples' Global Summit on Climate Change: 20-24 April 2009, Anchorage, Alaska* (Darwin, Australia: United Nations University—Traditional Knowledge Initiative, 2009)), pp. 5-7; available at http://www.indigenoussummit.com/servlet/content/home.html.

[202] This section prepared by Ronald O'Rourke, Specialist in Naval Affairs, Foreign Affairs, Defense, and Trade Division. It is adapted from CRS Report RL34391, *Coast Guard Polar Icebreaker Modernization: Background and Issues for Congress*, by Ronald O'Rourke.

[203] The Coast Guard's polar icebreakers also support U.S. scientific research in the Antarctic.

[204] *United States Coast Guard Arctic Strategy*, Washington, May 2013, p. 35, at http://www.uscg.mil/seniorleadership/ DOCS/CG_Arctic_Strategy.pdf.

[205] *United States Coast Guard High Latitude Region Mission Analysis Capstone Summary*, July 2010, pp. 12-13, 15.

[206] Source: Transcript of hearing.

[207] This section prepared by Ronald O'Rourke, Specialist in Naval Affairs, Foreign Affairs, Defense, and Trade Division.

[208] Dan Elliott, "NORAD Chiefs Eye Busier Arctic," *Fairbanks Daily News-Miner*, May 7, 2011.

[209] For a report assessing certain emergency scenarios in the Arctic, including search and rescue scenarios, see *Opening the Arctic Seas, Envisioning Disasters and Framing Solutions*, Coastal Response and Research Center, University of New Hampshire, report of January 2009, based on conference held March 18-20, 2008, at Durham, New Hampshire.

[210] Alister Doyle, "Arctic Nations Must Urgently Improve Rescue Services—Canadian Experts," Reuters (http://in.reuters.com), May 13, 2013.

[211] Cid Standifer, "Coast Guard Comandant: Service Still Committed To Eight NSCs," Inside the Navy, January 17, 2011.

[212] Jennifer Scholtes, "Papp: Coast Guard Arctic Plan Must Look Beyond Icebreakers," *CQ Homeland Security*, March 1, 2011.

[213] Source: Transcript of hearing.

[214] Source: State Department website, http://www.state July 7, 2011.

[215] The guidelines, as stated in Article 7, are as follows:

(a) search and rescue operations conducted pursuant to this Agreement in the territory of a Party shall be carried out consistent with the laws and regulations of that Party;

(b) if a search and rescue agency and/or RCC of a Party receives information that any person is, or appears to be, in distress, that Party shall take urgent steps to ensure that the necessary assistance is provided;

(c) any Party having reason to believe that a person, a vessel or other craft or aircraft is in a state of emergency in the area of another Party as set forth in paragraph 2 of the Annex shall forward as soon as possible all available information to the Party or Parties concerned;

(d) the search and rescue agency and/or RCC of a Party that has received information concerning a situation provided for in subparagraph (b) of this paragraph may request assistance from the other Parties;

(e) the Party to whom a request for assistance is submitted shall promptly decide on and inform the requesting Party whether or not it is in a position to render the assistance requested and shall promptly indicate the scope and the terms of the assistance that can be rendered;

(f) the Parties shall ensure that assistance be provided to any person in distress. They shall do so regardless of the nationality or status of such a person or the circumstances in which that person is found; and

(g) a Party shall promptly provide all relevant information regarding the search and rescue of any person to the consular or diplomatic authorities concerned.

[216] Article 20 identifies the government of Canada as the depository for the agreement.

[217] Source: Text of final version of agreement made ready for signing and dated April 21, 2011, http://arctic-council.org/filearchive/Arctic_SAR_Agreement_EN_FINAL_for_signature_21-Apr-2011.pdf.

[218] Source: State Department website, http://www.state July 7, 2011.

[219] This section was prepared by Carl Ek, who was a Specialist in International Relations, Foreign Affairs, Defense, and Trade Division, until his retirement on April 30, 2014. For questions relating to this section, contact Derek E. Mix, Analyst in European Affairs, Foreign Affairs, Defense, and Trade Division.

[220] "Arctic Sea Ice Continues to Decline, Hits Near Record Low," *IB Times*, October 4, 2011.

[221] "China Pushes for Arctic Foothold – From a Thousand Miles Away," Christian Science Monitor Global News Blog, March 7, 2013.

[222] "China, India Enter Heating-up Arctic Race," *Asia Times,* January 25, 2012; "Aston in Far North to Reach EU Policy on Arctic," *Agence France Presse,* March 8, 2012; "EU Application to Arctic Council Postponed," North Norway European Office, May 18, 2011 http://www.northnorway.org/nyheter/847-eu-application-to-arctic-councilpostponed See also: "Arctic Governance: Balancing Challenges and Development," European Parliament Regional Briefing 2012, DG EXPO/B/PolDep/Note/2012_136 June, 2012.

[223] Conference on the Arctic Ocean, May 26, 2008, Arctic Council website: http://arctic-council.org/article/2008/5/conference_on_the_arctic_ocean.

[224] Conference In Ilulissat, Greenland: Landmark Political Declaration on the Future of the Arctic, Edited December 11, 2008, website of the Danish Foreign Ministry, http://www.um.dk/en.

[225] A New Security Architecture for the Arctic: An American Perspective, By Heather A. Conley, Center for Strategic and International Studies, January, 2012, p. 14.

[226] "States Set Rules on Exploiting Arctic Wealth," *Agence France Presse*, May 12, 2011.

[227] U.S. Department of State, "Secretary Clinton Signs the Arctic Search and Rescue Agreement with Other Arctic Nations," May 12, 2011. "Warming Arctic Opens Way to Competition For Resources," *Washington Post*, May 16, 2011.

[228] "Clinton in Arctic to See Impact of Climate Change," *Agence France Presse,* June 2, 2012.

[229] For additional information, see Japan, Korea, Singapore and the Arctic Sea Lanes. *The Diplomat.* March, 2014.

[230] The European Union (EU) had applied for observer status, but its request was put on hold pending the resolution of a dispute with Canada and indigenous peoples regarding commercial sealing.

[231] "The Arctic Council: Underpinning Stability in the Arctic," The Arctic Institute/Centre for Circumpolar Security Studies, March 26, 2013, http://www.thearcticinstitute.org/2013/03/the-arctic-council-underpinning.html.

[232] "Nations Ponder How to Handle Busier, More Polluted Arctic," *Climatewire,* May 13, 2013, http://www.eenews.net/climatewire/stories/1059981040.

[233] Canada's Arctic Council Chairmanship. Ministry of Foreign Affairs, Trade, and Development Canada. Modified February 7, 2014. http://www.international.gc.ca/arctic-arctique/chairmanship-presidence.aspx?lang=eng.

[234] "The Arctic: Special Report," *The Economist,* June 16, 2012, p. 11.

[235] Canada's Arctic Passage Still a Ways Off. *Maritime Executive.* March 27, 2014.

[236] Transit Statistics. Northern Sea Route Information Office. Accessed April 20, 2014. http://www.arctic-lio.com/nsr_transits.

[237] During an October 2011 conference on Arctic shipping, President Putin enthused that "the Arctic is the shortcut between the largest markets of Europe and the Asia-Pacific region.... It is an excellent opportunity to optimize costs." "Warming Revives Dream of Sea Route in Russian Arctic," *New York Times,* October 17, 2011.

[238] Arctic LNG Project Backs Russia's Political Strategy. *Maritime Executive.* April 15, 2014.

[239] "Foreign Ministers of Russia and Canada Agreed To Go To the Ice Without the Third Parties," *WPS: Defense & Security,* September 20, 2010. "Russia Will Staunchly Defend its Interests in the Arctic Region – Putin," *Interfax.* June 30, 2011. "Russia to Deploy Troops to Defend Interests in Arctic," *Wall Street Journal*, July 2, 2011; "Russia Lets China Into Arctic Rush as Energy Giants Embrace," *Bloomberg.com,* March 25, 2013; "Russia, U.S., Norway to Hold Joint Naval Drills in August," *RIA Novosti,* May 24, 2012 http://en.rian.ru/world/20120524/173654802.html "Norway Hails Northern Eagle as Bridge-builder," *Defense New;* August 24, 2012; "China Eyes Arctic Options in Energy, Transport," *CNN Wire.* May 16, 2013.

[240] Naval Exercise on Hold Due to Ukraine Tensions. *Navy Times.* March 13, 2014. Hillary Warns Against Russia in Arctic. Barents Observer. April 3, 2014. http://barentsobserver.com/en/arctic/2014/04/hillary-warns-against-russiaarctic-03-04 Canada Takes Principled Stand on Arctic Council Meetings. Ministry of Foreign Affairs, Trade, and Development Canada. April 15, 2014. http://www.international.gc.ca/media 04/15a.aspx?lang=eng Crimea Crises Changes Nordic Defense Perspective.

[241] China's Arctic Play: Just the Tip of the Iceberg. *The Diplomat.* April 5, 2014.

[242] It was reported that China in December 2012 "offered Greenland a $5 billion investment package that included rights to mine iron, gold, nickel, and rare earth metals.... " See "Denmark, Norway Focus on Arctic Security," *Defense News,* March 11, 2013. China's Strategic Arctic Interests. *Strategic Comments.* IISS. Vol. 20, Comment 6. March 2014.

[243] "What Is China's Arctic Game Plan?" *The Atlantic,* May 2013.

[244] "China Plans First Commercial Trip through Arctic Shortcut in 2013," *Reuters.* March 12, 2013. "China Pushes for Arctic Foothold—From a Thousand Miles Away," Christian Science Monitor Global News Blog, March 7, 2013.

[245] "The Dragon Eyes the Top of the World," David Curtis Wright, Naval War College, China Maritime Studies Institute, Number 8, August 2011.

[246] "China Defines Itself as a 'Near-Arctic State,' Says SIPRI," Swedish International Peace Research Institute, May 10, 2012. "China's Arctic Ambitions," *Stratfor,* June 18, 2012. "China: Icebreaking in the Arctic," *The ISN Blog,* May 4, 2012; "Ice Station Dragon: China's Strategic Arctic Interest," *Defense News*, May 16, 2011.

[247] "China Knocks on Iceland's Door," *New York Times,* March 12, 2013; "China Signs Free Trade Agreement with Iceland as it Looks for Foothold in the Arctic," *The Washington Post,* April 16, 2013. China Willing To Expand Co-op with Iceland: Vice Premier. ECNS.CN. October 28, 2013. http://www.ecns.cn/2013/10-28/85976.shtml.

[248] "Chinese Icebreaker Concludes Arctic Expedition," *Xinhua News Agency,* September 27, 2012. Chinese Make First Successful North Sea Route Voyage. *The Arctic Journal.* September 12, 2013.

[249] White House, National Strategy for the Arctic Region, May 2013. http://www.whitehouse.gov/sites/default/files/docs/nat_arctic_strategy.pdf.

[250] U.S. Department of State, "Remarks at the Arctic Council Ministerial Session," U.S. Secretary of State John Kerry, available at http://www.state

[251] U.S. Department of State, "Kiruna Declaration: On the Occasion of the Eighth Ministerial Meeting of the Arctic Council," May 15, 2013, available at http://www.state

[252] Implementation Plan for the National Strategy for the Arctic Region. The White House. January, 2014. http://www.whitehouse.gov/sites/default/files/docs/implementation_plan_for_the_national_strategy_for_the_arctic_region_-_fi.... pdf.

[253] Adaptation of a letter from Senator Murkowski to President Obama. *Alaska Dispatch.* February 13, 2014. Secretary Kerry Announces Department Will Establish a Special Representative for the Arctic Region. U.S. Department of State. Press Release. February 14, 2014.

[254] Heather Conley, "The Colder War: U.S., Russia and Others Are Vying for Control of Santa's Back Yard," *Washington Post,* December 23, 2011.

[255] "Welcome To a New Era of Arctic Security," Rob Huebert, Canadian Defense and Foreign Affairs Institute, August, 2010.

[256] See, for example, "Canada Vows 'Firm' Defence of Arctic Border," *Canwest News Service,* November 24, 2009. "Danes With Dogs To Join Military Sovereignty Patrol Of Canadian Arctic," *The Canadian Press,* March 3, 2010. "Norway Plans One New Arctic Base, Perhaps More," *Defense News,* March 8, 2010.

[257] "The Arctic: Special Report," *The Economist,* June 16, 2012, p. 10.

[258] "Hands Across the Melting Ice," by James F. Collins, ross A. Virginia, and Kenneth Yalowitz, *New York Times,* May 13, 2013. "Oil Executives Tune Out the Call of the Wild Arctic," *Reuters,* May 31, 2013.

[259] "The Arctic Council: Underpinning Stability in the Arctic," the Arctic Institute/Centre for Circumpolar Security Studies, March 26, 2013. http://www.thearcticinstitute.org/2013/03/the-arctic-council-underpinning.html.

[260] "A New Security Architecture for the Arctic: An American Perspective," Center for Strategic and International Studies (CSIS), January 2012, p. 37.

[261] "Arctic Nations Meet to Discuss Communication, Maritime Domain Awareness Strategy," EUCOM website, August 30, 2012, http://www.eucom.mil/blog-post/24109/arctic-nations-meet-to-discuss-communication-

[262] "General Natynczyk and Fellow Northern Chiefs of Defence Discuss Shared Arctic Interests," News release, Canadian National Defense web page, April 13, 2012 http://www.forces.gc.ca/site/news-nouvelles/news-nouvelleseng.asp?id=4151 "Northcom Strives to Promote Safe, Secure Arctic," U.S. Department of Defense web page, December 17, 2012, http://www.defense.gov/news/newsarticle.aspx?id=118814.

[263] "A New Security Architecture for the Arctic: An American Perspective," Center for Strategic and International Studies (CSIS), January 2012, p. 30.

[264] "Russia, Norway Sign Border Deal For Arctic Energy," *Reuters*, September 15, 2010. "NATO Chief Cautions Against Division Over Arctic," *Canwest News Service*, January 29, 2009. "NATO Proposes Arctic Cooperation With Russia," Reuters, January 29, 2009. Speech by NATO Secretary General Jaap de Hoop Scheffer On Security Prospects In the High North, January 29, 2009, http://www.nato.int/docu/speech. "NATO Chief Wars Of Climate Change Security Risks," *Agence France Presse*, October 1, 2009. "Rogozin Says He Won't Discuss Cooperation In Arctic With NATO," *Interfax: Russia & CIS General Newswire,* March 27, 2009.

[265] "NATO Tests Its Forces in Arctic," February 16, 2012, Atlantic Council web page: http://www.acus.org/natosource/nato-tests-its-forces-arctic, "NATO, Russia Stage Arctic War Games," April 25, 2012. Atlantic Council web page: http://www.acus.org/print/68922. See also Cold Response 2014, Norwegian Armed Forces web page: http://mil.no/excercises/coldresponse/Pages/about.aspx. "Øvelse Cold Response er en norskledet, internasjonal øvelse som foregår i store deler av Nord-Norge," Norwegian Ministry of Defense, January 21, 2013, http://forsvaret.no/aktuelt/ovelser/Sider/Cold-Response.aspx.

[266] "Security in the High North: NATO's Role," NATO Parliamentary Assembly draft report, 071 PCTR 13 E, March 28, 2013.

[267] "NATO Won't Up Presence in the Arctic: Chief," *Agence France Presse,* May 8, 2013. "NATO Rejects Direct Arctic Presence," *Defense News,* May 27, 2013.

[268] Between a Rock and a Cold Place? NATO and the Arctic. January 15, 2014. Atlantic Council website: http://www.atlanticcouncil.org/blogs What the New NATO Secretary General Appointment Means for the Arctic. *The Fletcher Forum.* April 8, 2014.

[269] "Russia Will Protect Interests in Arctic: Official," *Agence France Presse,* June 10, 2009. "Russia Opposes Arms Race In Arctic Region—Diplomat," *ITAR-TASS World Service,* July 21, 2009.

[270] Putin Says Beefing Up Presence in the Arctic Among Top Priorities for Russian Military. *Associated Press.* December 10, 2013. Vows To Increase Russia's Arctic Military Presence. *The World Post.* December 10,

2013. The Arctic: The Arctic: Where the U.S. and Russia Could Square Off Next. *The Atlantic*. March 28, 2014.

[271] Nevertheless, Norwegian petroleum giant Statoil stated that it would continue with its joint drilling operations with Russia. Norway's Statoil Says Business as Usual in Russia. *Reuters News*. April 14, 2014.

[272] NATO's Relations with Russia. Accessed April 24, 2014. NATO website: http://www.nato.int/cps/en/natolive/topics_50090.htm, Begich: Military Presence in the Arctic Is 'More Important than Ever.' *The Hill*. March 19, 2014.

[273] "China Prepares For an Ice-free Arctic," Linda Jakobson, *SIPRI Insights On Peace and Security*, No. 2010/2, March 2010. "Admiral Urges Government to Stake Claim in the Arctic," *South China Morning Post*, March 6, 2010. "Norway Wants to Block China from Arctic Council Over 2010 Nobel Peace Prize Row," *Agence France Presse*, January 25, 2012. "Clinton in Arctic to See Impact of Climate Change," *Agence France Presse,* June 2, 2012.

[274] "China Pours Cash into Melting Arctic in Bid to Win Influence," *RIA-Oreanda News* [Russia], April 2, 2013.

[275] This section prepared by Ronald O'Rourke, Specialist in Naval Affairs, Foreign Affairs, Defense, and Trade Division.

[276] See, for example, Elias Torsson, "Washington's Folly," The Arctic Journal (http://arcticjournal.com), May 20, 2014.

[277] See, for example, Michelle Tan, "Inside the Army's Growing Arctic Circle Mission," Military Times (www.militarytimes.com), May 7, 2014.

[278] See, for example, Gwladys Fouche, "Wary of Russia, Norway Urges NATO Vigilance in Arctic," *Reuters.com*, May 20, 2014; Jon Mitchell, "Russia's Territorial Ambition and Increased Military Presence in the Arctic," *Foreign Policy Journal (www.foreignpolicyjournal.com)*, April 23, 2014; Atle Staalesen, "Hillary Warns Against Russia in Arctic," *Barents Observer (http://barentsobserver.com)*, April 3, 2014; Uri Friedman, "The Arctic: Where the U.S. and Russia Could Square Off Next," *The Atlantic (www.theatlantic.com)*, March 28, 2014; Julian E. Barnes, "Cold War Echoes Under the Arctic Ice," *Wall Street Journal (*http://online.wsj.com*)*, March 25, 2014.

[279] Department of Defense, *Quadrennial Defense Review Report*, February 2010, p. 86.

[280] Department of Defense, *Quadrennial Defense Review Report*, February 2010, p. 19.

[281] Department of Defense, *Quadrennial Defense Review Report*, February 2010, p. 57.

[282] Department of Defense, *Quadrennial Defense Review Report*, February 2010, p. 62.

[283] For an article discussing the change, see Jim Garamone, "Unified Command Plan Reflects Arctic's Importance," *American Forces Press Service*, April 7, 2011.

[284] The direction was contained on page 337 of H.Rept. 111-491 of May 21, 2010, the House Armed Services Committee's report on H.R. 5136, the FY2011 National Defense Authorization Act.

[285] Department of Defense, *Report to Congress on Arctic Operations and the Northwest Passage*, OUSD (Policy), May 2011. pp 2-4 (executive summary). See also Christopher J. Castelli, "Report: 'Significant' Gap Undermines DOD Tracking Of Ships In Arctic," *Inside the Pentagon*, June 16, 2011.

[286] The direction was contained on page 291 of H.Rept. 112-78 of May 17, 2011, the House Armed Services Committee's report on H.R. 1540, the FY2012 National Defense Authorization Act.

[287] Government Accountability Office, *Arctic Capabilities[:]DOD Addressed Many Specified Reporting Elements in Its 2011 Arctic Report but Should Take Steps to Meet Near- and Long-term Needs*, GAO-12-180, January 2012, Summary page.

[288] Department of Defense, *Arctic Strategy*, November 2013, 14 pp. The news release announcing the document is posted at http://www.defense.gov/Releases/Release.aspx?ReleaseID=16389. The document itself is posted at http://www.defense.gov/pubs/2013_Arctic_Strategy.pdf.

[289] Department of Defense, *Arctic Strategy*, November 2013, p. 2.

[290] Department of Defense, *Arctic Strategy*, November 2013, p. 3.

[291] Department of Defense, *Arctic Strategy*, November 2013, p. 7.

[292] Department of Defense, *Arctic Strategy*, November 2013, pp. 12-13.

[293] Department of Defense, *Arctic Strategy*, November 2013, p. 13.

[294] *Implementation Plan for The National Strategy for the Arctic Region*, January 2014, pp. 6-32.

[295] *Implementation Plan for The National Strategy for the Arctic Region*, January 2014, pp. 15-16.

[296] Christopher J. Castelli, "DOD, Canada Aim TO Develop Arctic Policy Options By Early 2010," *Inside the Navy*, December 7, 2009.

[297] Christopher J. Castelli, "American, Canadian Defense Officials Tighten Ties On Arctic Issues," *Inside the Navy*, May 3, 2010.

[298] NORAD news release, "NORAD, USNORTHCOM and Canadian Joint Operations Commmand Commander Sign Cooperative U.S./Canada Documents," December 11, 2012, accessed December 21, 2012, at http://www.norad.mil/ News/2012/121112a.html.

[299] Gerard O'Dwyer, "Arctic Nations Set Cooperation Guidelines," *DefenseNews.com*, June 27, 2013. See also Gerard O'Dwyer, "Greenland Meeting Highlights Arctic's Growing Importance," *DefenseNews.com*, June 2, 2013.

[300] David Pugliese, "Norad To Increase Focus On Arctic Surveillance," *Ottawa Citizen (www.ottawacitizen.com)*, March 3, 2014.

[301] David Pugliese, "Canada, US Eye Arctic Responsibilities for NORAD," *DefenseNews.com*, May 3, 2014.

[302] NSPD 66/HSPD 25, Section III B.

[303] Memorandum for Distribution dated November 10, 2009, from Admiral J. W. Greenert, Vice Chief of Naval Operations, on the subject of the Navy Arctic Roadmap, accessed July 24, 2013 at http://www.navy.mil/navydata/ documents/USN_artic_roadmap.pdf.

[304] U.S. Navy, *U.S. Navy Arctic Roadmap*, Washington, 2009, 29 pp. (October 2009, sponsored by Task Force Climate Change [and] Oceanographer of the Navy).

[305] Department of the Navy, *Arctic Environmental Assessment and Outlook Report*, August 2011, 25 pp., accessed July 24, 2013, at http://greenfleet.dodlive.mil/files/2011/08/U.S.-Navy-Arctic-Environmental-Assessment.pdf.

[306] Department of the Navy, *Arctic Environmental Assessment and Outlook Report*, August 2011, p. v.

[307] *Implementation Plan for The National Strategy for the Arctic Region*, January 2014, p. 23.

[308] *Implementation Plan for The National Strategy for the Arctic Region*, January 2014, pp. 6-32. The activity entitled "Develop a Framework of Observations and Modeling to Support Forecasting and Prediction of Sea Ice" is presented on pages 15-16.

[309] Department of Defense, *Arctic Strategy*, November 2013, p. 9.

[310] U.S. Navy, *U.S. Navy Arctic Roadmap 2014 – 2030*, February 2014, 43 pp., accessed February 26, 2014, at http://www.navy.mil/docs/USN_arctic_roadmap.pdf.

[311] U.S. Navy, *U.S. Navy Arctic Roadmap 2014 – 2030*, February 2014, pp. 3-4. Italics as in original.

[312] U.S. Navy, *U.S. Navy Arctic Roadmap 2014 – 2030*, February 2014, pp. 18-19.

[313] Steven Beardsley, "Navy Spots An Arctic Future, But Struggles To Plot A Course," *Stars and Stripes (www.stripes.com)*, March 24, 2014.

[314] David Grove, "Arctic Melt: Reopening a Naval Frontier?" *U.S. Naval Institute Proceedings*, February 2009: 16-21.

[315] David W. Titley and Courtney C. St. John, "Arctic Security Considerations and the U.S. Navy's Roadmap for the Arctic," *Naval War College Review*, Spring 2010: 41.

[316] David W. Titley and Courtney C. St. John, "Arctic Security Considerations and the U.S. Navy's Roadmap for the Arctic," *Naval War College Review*, Spring 2010: 42.

[317] Source: Text of remarks of Chief of Naval Operations Admiral Gary Roughead at Active in the Arctic Seminar, June 16, 2011.

[318] Source: Text of remarks of Chief of Naval Operations Admiral Gary Roughead at the 4th Symposium on the Impacts of an Ice-Diminishing Arctic on Naval & Maritime Operations, June 20, 2011. See also Christopher J. Castelli, "Roughead: International Approach Needed To Fix Arctic-Awareness Gap," *Inside the Navy*, June 27, 2011.

[319] Nicole Klauss, "War Games Find U.S. Navy Ability Lacking In Arctic," *Fairbanks Daily News-Miner*, April 28, 2012.

[320] For additional discussion, see "Polar Icebreaking".

[321] *United States Coast Guard High Latitude Region Mission Analysis Capstone Summary*, July 2010, pp. 10-11, 13-15.

[322] *United States Coast Guard Arctic Strategy*, Washington, May 2013, 47 pp.; accessed May 24, 2013, at http://www.uscg.mil/seniorleadership/DOCS/CG_Arctic_Strategy.pdf.

[323] *United States Coast Guard Arctic Strategy*, Washington, May 2013, pp. 9-10. Emphasis as in original.

[324] Antonieta Rico, "Papp: No Plans For More Coasties in Arctic," *NavyTimes.com*, June 1, 2013.

[325] H.R. 3619/P.L. 111-281 of October 15, 2010 is the Coast Guard Authorization Act of 2010. Section 307 concerned implementation of the Arctic Marine Shipping Assessment. Subsection 307(b) stated that "To carry out the purpose of this section, the Secretary of the department in which the Coast Guard is operating is encouraged to enter into negotiations through the International Maritime Organization to conclude and execute agreements to promote coordinated action among the United States, Russia, Canada, Iceland, Norway, and Denmark and other seafaring and Arctic nations to ensure, in the Arctic—(1) placement and maintenance of

aids to navigation; (2) appropriate marine safety, tug, and salvage capabilities; (3) oil spill prevention and response capability; (4) maritime domain awareness, including long-range vessel tracking; and (5) search and rescue.

[326] *Implementation Plan for The National Strategy for the Arctic Region*, January 2014, pp. 7-31.

[327] *Implementation Plan for The National Strategy for the Arctic Region*, January 2014, pp. 7-8, 13, 24, 25, 31, and 31, respectively.

[328] *Implementation Plan for The National Strategy for the Arctic Region*, January 2014, p. 13.

[329] CRS communication with State Department official, October 8, 2010.

[330] On this question, a January 19, 2009, press article about the new directive stated:

The new policy directive covers several key areas, including national security, energy exploration and the environment, but it does not specify whether any should take precedence over others.

That led Jeremy Rabkin, a professor at George Mason University Law School, to comment: "It's really a list of all the things we're concerned about; that's not policy. I don't see anything here that helps you decide what gets priority." (Juliet Eilperin and Spencer S. Hsu, "White House Directive Guides Policy On Arctic," *Washington Post*, January 19, 2009: 2.)

[331] Heather A. Conley et al, *The New Foreign Policy Frontier[:] U.S. Interests and Actors in the Arctic*, Washington, Center for Strategic & International Studies, March 2013, pp. 21-22.

[332] For examples of congressional reactions to the January 2014 implementation plan, see, the January 30, 2014, press release from the office of Senator Mark Begich entitled "Begich Calls on White House to Fund Arctic Plan," accessed February 26, 2014, at http://www.begich.senate.gov/public/index.cfm/2014/1/begich-calls-on-white-house-to-fundarctic-plan; and the February 13, 2014, press release from the office of Senator Lisa Murkowski entitled "Murkowski Slams "Unambitious" White House Arctic Plan," accessed February 26, 2014, at http://www.murkowski.senate.gov/ public/index.cfm?p=PressReleases&ContentRecord_id=30e9f7c2-28ce-4baf-9ffe-5171ac6ceaa8&ContentType_id= b94acc28-404a-4fc6-b143-a9e15bf92da4&Group_id=c01df158-d935-4d7a-895d-f694ddf41624. The February 13, 2014, press release cites a February 11, 2014, letter from Senator Murkowski to President Obama, which was accessed February 26, 2014, at http://www.murkowski.senate.gov/public/?a=Files.Serve&File_id=875ea38b-f881-459d-a2b1-220e17f99dc7.

[333] Title II of P.L. 98-373 is the National Critical Materials Act of 1984.

[334] Source for text: http://fas.org/irp/offdocs/nspd/nspd-66.htm. The text is also available at http://www.nsf.gov/od/opp/ opp_advisory/briefings/may2009/nspd66_hspd25.pdf.

[335] *National Strategy for the Arctic Region*, May 2013, 11 pp.; accessed May 14, 2013, at http://www.whitehouse.gov/ sites/default/files/docs/nat_arctic_strategy.pdf. The document includes a cover letter from President Obama dated May 10, 2013.

[336] A footnote in the document at this point states: "Arctic state is defined as one of the eight nations making up the permanent membership of the Arctic Council and includes the following nations: Canada, Denmark (including Greenland and the Faroe Islands), Finland, Iceland, Norway, Russia, Sweden, and the United States."

[337] A footnote in the document at this point states: "U.S. Geological Survey, "Circum-Arctic Resource Appraisal: Estimates of Undiscovered Oil and Gas North of the Arctic Circle," fact sheet, 2008, http://pubs.usgs.gov/fs/2008/ 3049/fs2008-3049.pdf."

[338] A footnote in the document at this point states: "National Security Presidential Directive-66/Homeland Security Presidential Directive-25, January 2009."

[339] A footnote in the document at this point states: "Much of this work is already underway including efforts under Executive Order 12501 (Arctic Research), Executive Order 13547 (Stewardship of the Ocean, Our Coasts, and the Great Lakes), and Executive Order 13580 (Interagency Working Group on Coordination of Domestic Energy Development and Permitting in Alaska). Entities under these Executive Orders are developing partnerships with Federal, state, local, tribal, territorial, public and private sector partners to ensure that natural resource decisions in the Arctic integrate economic, environmental, and cultural interests of the Nation."

The above-mentioned Executive Order 12501, which was issued January 28, 1985, implements The Arctic Research and Policy Act (ARPA) of 1984 (Title I of P.L. 98-373 of July 31, 1984) by, among other things, establishing the Arctic Research Commission and the Interagency Arctic Research Policy Committee. The text of Executive Order 12501 is available at http://www.archives.gov/federal-register/codification/executive-order/12501.html.

[340] A footnote in the document at this point states: "See Executive Order 13175 – Consultation and Coordination with Indian Tribal Governments, November 2000."

[341] A footnote in the document at this point states: "Alaska Federation of Natives (AFN) (1996) *AFN Board Adopts Policy Guidelines for Research*. AFN, Anchorage. Alaska Natives Knowledge Network. Policy adopted in 1993, http://ankn.uaf.edu/IKS/afnguide.html."

[342] *National Strategy for the Arctic Region*, May 2013, pp. 4-11.

INDEX

F

G

H

I

J

K

L

M

N

O

P

R

S

T

U

V

W